Marshall Islands

180°

Kiribati

Nauru

Phoenix Islands

Tuvalu

Tokelau

equator

Samoa

Cook Islands
(northern)

Vanuatu

Fiji

American Samoa

Tonga

Niue

Cook Islands
(southern)

French Polynesia

Kermadec Is.

SOUTH PACIFIC OCEAN

New Zealand
Exclusive Economic Zone
(EEZ)

New Zealand

Chatham Is.

Bounty Is.

Antipodes Is.

Auckland Is.

Campbell Is.

135° W

180°

SOUTHERN OCEAN

45° S

ROSS SEA

THE CATCH

ALSO BY MICHAEL FIELD

Mau: Samoa's Struggle against New Zealand Oppression
Speight of Violence: Inside Fiji's 2000 Coup
Black Saturday: New Zealand's Tragic Blunders in Samoa
Swimming with Sharks: Tales from the South Pacific Frontline

THE
CATCH

How fishing companies reinvented slavery
and plunder the oceans

MICHAEL FIELD

AWA PRESS

First edition published in 2014 by Awa Press,
Unit 1, Level 3, 11 Vivian Street, Wellington 6011, New Zealand.

ISBN 978-1-927249-02-4

ebook formats
epub 978-1-927249-03-1
mobi 978-1-927249-04-8

A catalogue record for this book is available from the
National Library of New Zealand.

Cover photograph © Greenpeace / Roger Grace
Cover design by Greg Simpson
Typesetting by Tina Delceg
This book is typeset in Minion Pro and Founders Grotesk
Printed by Midas Printing International Ltd, China

Find more great books at awapress.com.

Produced with the assistance of

MICHAEL FIELD has been a newspaper and agency reporter for 42 years, mostly covering the South Pacific. A former correspondent for Agence France-Presse, he now reports for Fairfax Media and is Pacific affairs commentator on Radio New Zealand's Nine to Noon programme. As a Harkness Fellow he attended the John F. Kennedy School of Government at Harvard University. He has reported on the Pacific's fishing world – and its byzantine politics – from remote atolls to aboard a Soviet-era trawler.

Karakia hi tuna, hi ika hoki
Taku aho nei, ka tangi wiwini.
Taku aho nei, ka tangi wawana.
Taku aho nei, ka hinga, ka mate ra.
Kai mai, kai mai, e te kōkopu,
Ki taku nei mōunu nei.
Tara wiwini, tara wawana,
Kia ai he whakataunga mau
Ki te uru ti, ki te makau.
Tara wiwini, tara wawana,
E tuapeka ki Wai-korire.

Prayer used in catching eels or fish
My fishing line here, its cry is awesome.
My fishing line here, its cry is dreadful.
My fishing line here, it is dropped, it sinks there.
Bite here, bite here, O you kōkopu,
Bite here at my bait.
Awesome point, dreadful point,
That allows you to be landed
By the sharp point, by the hook.
Awesome point, dreadful point,
Which will deceive at Wai-korire.

*To the thousands of nameless people who brave the seas to feed us,
to their families who anxiously await their return,
and to those who grieve for the ones who never come home.*

A definition of modern slavery

A person is forced to work, held through fraud, under threat of violence, for no pay beyond subsistence. He or she cannot walk away from their job without dire consequences to them or their families.

'I was a slave, but then I became useless to the Koreans so they sent me home with nothing'

> – Ruslan, an Indonesian crewman
> on fishing boat *Melilla 203*

Contents

Author's note

Since the middle of the 20th century there have been attempts to impose order and control over fishing, and the exploitation of other resources, in the world's seas and oceans.

Traditionally a country's sovereign territorial waters extend 12 nautical miles – 22.22 kilometres – from its coastline. Starting with the United States in 1945, various countries unilaterally declared they held rights over waters beyond this, but it was not until 1982, with the passing of the United Nations Convention on the Law of the Sea, that extended rights were formalised. Under the convention, which entered into force in 1994, each country has rights over the waters 200 nautical miles – 370 kilometres – out from the edge of its territorial waters. As well as having the sole right to exploit natural resources within this exclusive economic zone, the country must protect the marine environment.

New Zealand declared its exclusive economic zone in 1978. The sixth largest in the world, it covers nearly four million square kilometres. With vast oceans on all sides, this should comprise a rich fishery. Instead, it has been poorly understood and vulnerable to overfishing.

In 1986, after finding its inshore fishery had too many boats and not enough fish, the New Zealand government created a quota management system. The fishing industry boasts that this has made New Zealand a world leader in sustainable fisheries management.[1] Almost all commercially targeted fish species within New Zealand's waters are now managed within the system.

The quota management system is complicated. Stripped down, it is designed to ensure there will always be enough fish to catch. Scientific formulas are used to calculate how much of a particular species in an area can be taken and still leave enough fish for the species to breed and replenish itself. Whether this works depends on a combination of science, research, and honesty – fishing boat skippers need to declare what they have caught. The beauty of the system is that, if a stock goes into decline, the quota can be cut.

Each season, scientists, in consultation with the industry, recommend to the government a total allowable commercial catch, or TACC, by volume (usually tonnes) for each of the 100 species covered by the system. This, in turn, is divided up into annual catch entitlement – ACE – which is the right to take a certain volume by weight of fish stock during the fishing year. In 2014 the 100 species (or species groupings) covered were divided into 636 separate stocks, with each being managed independently.[2]

As the system is based on individual 'property' rights, it gets complex. An individual transferable quota is the right to harvest a defined amount of a species – a percentage by weight of the TACC – in a specified area during a fishing year. If someone holds quota for six percent of the TACC for a particular species in a particular area, they hold the right to harvest six percent of the fishing area's TACC. The Fisheries Act limits how much quota any one person or company can own – so that no one company or individual can develop a monopoly.[3]

Heavy maths are needed to work out what quota converts into ACE each year: the government provides an online tool to work it out.[4]

Quota is divided between commercial fishermen and Māori. With some species such as snapper, New Zealand's most popular fish, there is also a catch set aside for recreational fishing.

There are numerous offences within the quota management system; they range from taking more fish than your entitlement to assorted accountancy and bookkeeping offences. Fishermen who misfile paperwork can pay heavily.

A quota is like other forms of property – it can be leased, bought, sold or transferred. For example, if an individual or company does not have enough annual catch entitlement to make a fishing mission worthwhile, they can lease out their entitlement to others who can consolidate enough quota to warrant a fishing trip.

Only New Zealanders and New Zealand-owned companies can own fishing quota in New Zealand, and foreign ownership of shares in New Zealand quota-owning companies is limited to 25 percent.[5] However, and this is at the core of this book, quota holders have been able to lease foreign vessels to catch their allowance on their behalf.

Preface

I felt peckish. It had been a long ride in a car with a decidedly surly driver and little to eat. Now there was time to kill before my flight. The food hall at Indira Gandhi International Airport in New Delhi offered the full range of Indian fast foods but they were a bit hefty for the occasion. I wanted something light. The ubiquitous McDonald's offered Filet-O-Fish. It would fit the bill in the ten minutes before the Mumbai flight boarded.

My selection was not unusual: fish makes up nearly 16 percent of the total human intake of animal protein. On average, every person on the planet eats 17 kilograms of fish a year. Globally, fish provides more than one and a half billion people with almost 20 percent of their average per capita intake of animal protein, and three billion people with at least 15 percent. In short, fishing is mighty important to the human race.[1]

Modern-day fishing in the oceans is like Buffalo Bill blasting away at bison in the American West: the supply seems endless but is in fact diminishing towards the point of extinction for many key species. Over 140 million tonnes of fish are taken each year, of which around 115 million tonnes go directly to human consumption and the rest to industrial processes creating food products, some of it to feed other fish.

The supply chain that had delivered the fish for my Filet-O-Fish to an airport in India was long and convoluted. Few McDonald's customers, having a snack while waiting for a flight, could hope to know (if they were even interested) who had caught the fish, much less how it had been processed. It is also highly unlikely they would know anything about the species or whether the fish had been sustainably harvested, so it would be impossible for them to make an ethical or considered choice. All of us would be entirely dependent on the moral considerations, if any, that McDonald's may have taken account of.

The fine print on the box said the fish was hoki. The fish's Latin name, *Macruronus novaezelandiae*, was not included, but had it been it would have revealed the fish's origins. This barely known species of fish, processed and sold in vast quantities around the world, is, as this book will show, hauled out of cold waters off New Zealand's South Island, sometimes by underpaid or barely paid men from poor areas of Asia.

Wild fishing is labour-intensive. Everything on a vessel, from shooting the net to processing the fish and packing it, requires human hands. After a fish hits the pound below the trawling deck, lines of men quickly cut off its head and tail, gut the rest, pack and snap-freeze it. These men may have been on the job for longer than any factory worker on land would be. They will be cold and wet. Some will be seasick, and all will be tired and sore. Their workplace will be crowded and noisy; depending on the seas, it will also move violently and without warning, throwing them into walls, into each other, and sometimes even into machinery used to process the fish.

They will also be expected to paint and maintain the boat, not an easy job for it will probably be an ageing rust bucket. Their meals may consist of damaged hoki. On some fishing boats even fresh damaged fish will be a luxury. In 2013 in Whangarei I came across Spanish fishing boats whose Indonesian crew had been fed 'bait fish' – frozen mackerel and squid – for 60 days straight. Meanwhile, the officers on the boat had been eating wholesome Spanish meals.

At some point the frozen hoki will be shipped to China, South Korea or Thailand. There it will be semi-thawed so that further cheap labour can cut it down into fillets – or cellphone-sized pieces to stick inside buns. No longer really fish at that point, it has become the end point of a supply chain that often relies, for its economic viability, on paying workers as little as possible, and sometimes nothing at all.

Then there is the other question: has the hoki been caught sustainably? Or is it destined to go the way of many other fish species?[2]

A couple of months later, I attend a family celebration at an eye-wateringly expensive Japanese restaurant in Auckland. The restaurant's set menu is designed to make pescetarians happy. Environmentalists would be less so. One course offers Antarctic toothfish from the Ross Sea. Eating Antarctic toothfish, one of the most expensive fish sold anywhere, is the maritime equivalent of eating tiger. This is a fish that no one has ever needed in order to sustain life. Only recently identified, it may, however, be required to stay in the ocean to maintain the balance of life on the planet.

The toothfish was delicious, the best-tasting fish I have come across. But should it be on the market at all? Twenty years ago only hardy polar scientists even knew this creature existed; few realised it was fleshy and tasty. Even as it is being harvested, little of its ecology is known. We are aware it plays a key role in the Antarctic ecosystem as a premier predator, but we know almost nothing of its breeding. By the time we find out, fishing may have done irreversible damage to the species. And, as I was to discover, toothfish is taken in ancient boats by enslaved crews who are dying in the Southern Ocean at an alarming rate.

The other fish on the menu, bluefin tuna, was even more ethically troubling: environmentalists believe it should be classified as an endangered species and not fished at all.

Increasingly, our food products are globalised, yet most of us know little about how they get to us. For some, this could be a happy circumstance. When I was a child, my farmer grandfather took me on a visit to the local meatworks. I was shown the 'Judas sheep' but luckily spared the outcome. I was to get the full details later as a young agricultural journalist in Botswana, doing a story about 'captive bolts' – pistols that fire pistons into animals' heads to stun them without the need for bullets – at the Lobatse abattoir. At the end of the tour the butcher down below handed over a couple of freshly made 'cheerios'.

Most of us eat meals that include meat, but very little meat, even venison, is caught in the wild any more. Instead, it is farmed. Fish are different. Most are gathered by nautical hunters and gatherers, harvesting from a wild stock that they rarely understand.

As this book was being written, data from the Food and Agricultural Organization of the United Nations was showing farmed aquaculture fish on the verge of overtaking wild-caught fish in volume. At first glance this seems a hopeful sign that wild fish stocks can be preserved. But farmed fish have to be fed something, and mostly they are fed wild fish.

On May 2, 2014, E. Robert Kinney died, aged 96. In the 1950s, as the boss of General Mills in the United States, Kinney created the fish finger – or the fish stick, as his company called it then. 'Thanks to fish sticks,' an advertisement proclaimed, 'the average American homemaker no longer considers serving fish a drudgery. Instead, she regards it as a pleasure … easy to prepare, thrifty to serve, and delicious to eat.'

In 1954 *The New York Times* reported that fish sticks were 'the newest best-seller in supermarkets. … Mothers report youngsters gobble the sticks like candy – or almost. … Their crisp surface and the fact they may be eaten with one's fingers attract children.' But, the paper added presciently, 'processing robs the fish of its identity and cloaks it in what apparently is an appetising anonymity.'[3]

Can the world, with a population now topping seven billion, afford to continue hunting fish in the way Americans hunted bison? Can we continue to see fish as just another commodity? I am not anti-fishing: there are enough fish in the ocean to provide the world's humans with the protein we need. But the harvesting of fish has to be done properly, with care, and with respect for human dignity.

1 Dangerous seas

As deep-sea fishing boats go, *Tai Ching 21* was nothing exceptional, just another battered, hard-working predator pursuing the world's wild fish across lonely oceans. Often the crew had only a vague idea of where they were, being too tired even to notice. Flying Taiwan's flag, in November 2008 the 18-year-old vessel had 29 men on board: the Taiwanese captain and 18 Chinese, six Indonesian and four Filipino crew members. The crew were on the barest of wages, and working in conditions close to a modern equivalent of slavery.[1]

The ship was chasing yellowfin tuna, fast, strong fish capable of great migrations requiring considerable endurance. Yellowfin tuna are attractive. The backs of their streamlined bodies are metallic blue and their stomachs a mix of yellow and gold. While not the most lucrative of the tuna species, they command good prices on the Japanese sashimi market. Sashimi, usually the first course in a Japanese meal, comprises thinly sliced raw tuna served with a sauce. It is an ancient dish, and an energy-efficient and non-wasteful way of consuming fish.

As nets damage the catch, fish for sashimi needs to be caught by hook. For that reason, *Tai Ching 21* was a longliner. It set out lines of up to 100 kilometres; each had up to 3,000 hooks, aiming to catch fish of around 25 kilograms.

In essence, a longliner finds a school of fish and then, using boat speed, hauls a line through it. Careful skippers will avoid also killing sharks, billfish such as marlin and swordfish, sea turtles and seabirds in the process. Others do not care.

Tuna migrate across the Pacific Ocean each year in a never-ceasing movement in pursuit of food and ahead of that which would eat them. They pass through the exclusive economic zones of numerous Pacific states, as well as areas of the world once quaintly called 'high seas' – global commons owned by no one and open to pillage by anyone with the desire to do so.

Fish masters no longer smell the air and guess where yellowfin are. With the help of technology, they look for areas in the ocean where life is richer than usual. Various currents and eddies are signposts, picked up by satellites measuring ocean temperatures and spotting chlorophyll blooms. (Many people flying in a commercial airliner over an ocean will see blooms and waters of a different temperature without recognising what they are.) And, too, the ocean floor is increasingly being mapped, including for undersea mining. Particular features, notably seamounts – mountains that rise hundreds of metres from the seabed – can influence which fish are near the surface.

For the crew of a longliner, though, details of how the fish are discovered are academic: a big catch just means a great deal more work for little reward. For the men on *Tai Ching 21,* the work was brutal. Large live fish would come on the deck still hooked; getting them off the hook and into the hold with as little damage as possible was physically tough. And once a fishing boat – trawler or longliner alike – is taking its catch, there are no breaks. As long as fish are coming aboard, the work keeps going until the holds are full or the fish move on.

On October 28, 2008, the captain of *Tai Ching 21* used a satellite phone to catch up with his wife in Taipei. He was fishing around Kiribati's Phoenix Islands, eight near-pristine coral atolls that were part of one of the world's largest marine protected areas. The protection was a diplomatic lie:

fishing in various guises by many nations took place, but Kiribati could claim to be innocent and unknowing. One of the poorest nations in the Pacific and among its most desperate, it allowed fishing in its marine protected area by favoured nations. Tuna licensing was too lucrative to let the fish swim on to another nation's waters to be caught there.[2]

The phone call to Taipei was the last anyone heard from *Tai Ching 21*. After this there was only silence. There was not even a digital alert from the boat's EPIRB, its automatically triggered emergency position-indicating radio beacon.

On November 9 a Korean fishing vessel came alongside *Tai Ching 21*. The Taiwanese ship was afloat and in no danger of sinking. There had been a fire. Three life rafts and a lifeboat were missing, as were all 29 men.

What followed was one of the most extensive Pacific search operations in recent times. As is typical with such events, the boat's flag state, in this case Taiwan, paid not a cent. Nor did the boat's owner make any kind of contribution to finding the men it had used to make profit on the high seas. A Royal New Zealand Air Force P3 Orion searched 54,000 square kilometres, an area roughly one and a half times the size of Taiwan itself. The United States Air Force put up a C130 Hercules. Nothing was found.

The last time there had been such a big search in the area was July 1937, when the Americans were looking for lost aviators Amelia Earhart and Fred Noonan. Even in 2012, people were still looking for Earhart and Noonan. After a month, no one was any longer looking for the men of *Tai Ching 21*.

As a reporter, I know that names make stories: you can put much of a story together with a couple of names, but get names wrong and you will not hear the last of it. I was aware that when it came to fishing boat stories, names were a rarity. It was a quietly understated piece of racism – the names would inevitably be Asian, and for Western reporters it could be hard to pick what part of a name was the surname, or whether the name was the right way around.

In my own time I made an effort to put names to the disappeared crew of *Tai Ching 21*. Maritime unions, police, churches and embassies all told me they did not know. The Royal New Zealand Air Force had no idea whom they had spent so much time trying to find. I gave up without finding a single name, but my search had opened a window into a world where men no longer had names, where people who sold the men's labour for a cut controlled their identities.

Tai Ching 21 was towed back to Taipei, repaired and put back to sea. The price of yellowfin tuna on the Tokyo market may have registered a blip because of the lost cargo, but there is no shortage of boats chasing tuna these days. *Tai Ching 21* was little more than a routine newspaper story without much of an ending, except that still today there are places across Asia where families quietly mourn their kin who put to sea in a bid to feed their families and never came home.

2 Frozen to death

Move around the fishing industry or anything to do with it and you will hear a lot about how dangerous it is. Management and owners, as a way of trying to proactively absolve themselves from responsibility for the tide of death and injuries, often say it themselves. At the same time, there's an unspoken implication that people who don't go to sea – who buy their fish rather than catch it – have no right to question the industry.

It is true that deep-sea fishing has always been an occupation whose workers face a limited life expectancy. Those who fished Atlantic cod from the sixteenth century onwards faced appalling risks, if only because the sail-powered boats, and later engine-powered vessels, could barely cope with the seas, let alone the large fish they struggled to haul aboard. But the industry today has little excuse for the fatalities and injuries that occur. People are being killed not as an inevitable consequence of the battle between man and ocean, but because owners take shortcuts and gamble with the lives of men who have few options.

This is especially so in the pursuit of Southern Ocean toothfish, of which there are two species – the Antarctic, or Ross Sea, toothfish and the Patagonian toothfish – both sold as 'Chilean sea bass' in United States markets keen on fudging the truth.

Toothfishing is not about chasing protein to feed the world's hungry masses. The fish are sold at massive profit to a few selected buyers at luxury resorts in Las Vegas and top restaurants in New York and Tokyo (and in some places in Auckland, New Zealand). Only the rich can afford to eat toothfish.

For Korean companies such as Insung Corporation, the only concern when toothfishing in the Southern Ocean was filling the quota they had been granted by the Commission for the Conservation of Antarctic Marine Living Resources. When CCAMLR had come into existence in 1980 it had been little more than a sideshow to the Antarctic Treaty, which governed the scientific operations in, and access to, Antarctica. There was virtually no fishing; the commission was primarily founded to control what was seen at the time as a large take of krill, a tiny creature at the bottom of the food chain. Scientists worried that depletion of krill would impact on the rest of the ecosystem.

Studies commissioned by CCAMLR and those of other industry groups give the impression that Antarctic krill is an almost limitless resource: the commission's current annual catch limit is 620,000 tonnies. But given that this creature is essential food for much Antarctic wildlife, from baleen whales to penguins, it seems extravagant and dangerous to be industrially fishing it until the overall impact of this is known. Given the remoteness of the fishery and its attractiveness to factory boats now being forced out of other fisheries, krill is a tempting take beyond the reach of law enforcement.

The beginning of the toothfishing industry in the Ross Sea and south of the Atlantic and Indian Oceans saw the commission increase its oversight and control of who could fish, and for what. Its secretariat, based in Hobart, Australia, issued quotas and rules to member nations. However, it had little power of enforcement. Its annual meetings were attended mainly by scientists and fishing industry representatives and attracted little media attention.

This changed as the industry expanded and governments began to realise the great fishing potential of the Southern Ocean. Diplomats and politicians now attend the commission's meetings, and a blossoming roll call of nations has signed up. As at December 2013 there were 25 member states. But while the commission sets the annual quotas for krill and toothfish, it has proven indifferent to the suitability or otherwise of the boats that take them.

Going after toothfish, a creature that can weigh over 140 kilograms and be up to two metres long, is deemed by the industry to be 'Olympic fishing' because of the extreme conditions and the limited fishing season. Late in 2010, flying the flag of South Korea, *Insung 1* sailed out of Montevideo, Uruguay, and headed into the Southern Ocean, around Cape Horn, and towards the Ross Sea due south of New Zealand.

Insung 1 was 31 years old and little suited to its environment. Like many boats involved in deep-sea fishing, it had carried several names and flags. Built in 1979 for a Japanese company, it had been launched as *707 Bonanza*. In 2000 it had nearly sunk when a large amount of sea water had flowed in through the trawling door; another ship with pumps had been needed to save it.

In 2005 the Japanese sold the boat to Insung Corporation. According to the company, the previous owners didn't tell them about the accident.

The boat had been built for tropical waters. Insung modified it, but only in a limited way, so it could go into the Southern Ocean. It erected light steel bulkheads on the trawling deck to offer the crew sufficient shelter from the sea that they could continue working when it got rough. There was no attempt at ice strengthening.

Insung Corporation operates 14 fishing boats. Each has on board a set of *The Regulation for the Safety of Fishing Vessel*s, but only in Korean. *Insung 1* had 40 officers and crew and two Korean observers. The 33 crew members were Chinese, Indonesian, Vietnamese and Filipino. All had been recruited from six hiring agencies: Panworld, the Korea Overseas Fisheries Association, Hwa Young Trading, Hanjin Group, Changman and Noah International. The Chinese sailors, who were of Korean descent and probably North Korean refugees, were able to speak Korean, but not English. The Filipino men spoke English, but not Korean. The Indonesians spoke limited English and no Korean, and the Vietnamese spoke neither English nor Korean.

Indonesian workers on fishing boats are especially vulnerable. Most come from Tegal in central Java, an area marked by poverty and high unemployment. Offers of work paying between US$250 and $500 a month are attractive to people living below a poverty line of around US$75 a month. To get the jobs, the men have to pay a cash bond of around US$300 and hand over birth certificates and family documents – in fact, any documents that prove their existence and identity.

Insung 1 began fishing in the Southern Ocean on December 1 and within nine days had taken 39 tonnes of toothfish. It and fellow Korean boats *Hong Jin 707*, *Jung Woo 2* and *Jung Woo 3* had a combined quota of 372 tonnes. *Insung 1*'s officers now wanted to move to a new zone, over an area known as Trench 88-1.

By now the ship was dodging ice floes frequently. On December 13 it faced typical Southern Ocean conditions. The sky was overcast with visibility of around eleven kilometres, the wind was blowing at 37 to 40 kilometres an hour, and there was a three-metre swell. Although any ship would roll and pitch in such conditions, they were not extreme. The ride would not be comfortable but a well-found ship and a warm and well-cared-for crew would cope.

Everything changed with a decision by the 45-year-old captain, Yu Yeong-seob, to leave open the trawl door at the stern. Perhaps he always did this, or perhaps he just didn't think about it. He may have got away with it but another of his decisions, to change course, would prove disastrous. It put him with a following wind.

Sea water sloshed through the open trawl door and flooded into a passageway used for storing fishing gear. The passageway was slightly sloped toward the bow so water would flow down towards an electric pump, which would automatically move the water overboard. The pump was not working.

The boat started listing to starboard. Water continued to flow in and move through the rest of the ship, soon reaching the engine room. Yu told the men in the engine room to transfer fuel to the port-side tanks, and ordered the crew to move fishing sinkers and gear to the port side.

An officer who had been asleep woke at around 6.20 a.m. and realised the ship was on a dangerous list. He raced to the bridge and told Yu, who ordered the bow to face the leeside. Yu then sent a message to the nearby *Hong Jin 707*: 'The ship is about to be overturned.' Bridge alarms sounded and the main engine stopped. Yu ordered the crew to the bridge and a deckhand handed out life jackets. At 6.25 a.m. the ship capsized and sank. It was 1,850 kilometres north of Scott Base.

Two inflatable rafts attached to the ship automatically released and inflated, but they landed about ten metres from the vessel. Crew members jumped into the freezing water, many getting caught up in reels and buoy lines that were floating on the surface. There had been no safety training and there was no proper safety gear. The men were wearing life jackets, but not survival suits that could have protected them from the icy waters until help arrived. Survivors would later tell of panic. Most of the men must have realised as the ship slid into the water that they were doomed. Death would have been quick; 22, including Captain Yu, died, most of them from hypothermia.

The Korean Maritime Safety Tribunal duly investigated the sinking. Its report skirted around the open doors. It blamed the crew for the disaster, although it did suggest language differences had played a part. 'While one should not discount the practical difficulties of recruiting and hiring foreign sailors for pelagic fishing vessels, it is also difficult to avoid the conclusion that the ship owner should have made greater efforts and precautions to ensure safety in fishing and emergency situations alike, given the risks implicit in the significant language barrier that existed among the crew members.'[1]

In hindsight the captain should have ensured the doors were closed, the tribunal said. It was 'the neglect of the doors [that] led to a fatal situation when the ship changed its course... The rolling of the hull, and the increasing influx of sea water through the trawling door on to the upper deck as well as into the passageway for fishing gear, eventually sent the ship down to the bottom of the ocean.'

The captain's various attempts to save the situation had been inept and at no point had he tried to either close doors or check whether pumps were working. 'Upon noticing that the ship's hull was radically tilting to starboard, the captain should also have set off the emergency alarms, gathered all crew members on the designated evacuation spot, and evacuated the ship accordingly. The captain's failure to command evacuation in a timely manner prevented the crew members from floating the rescue boats before the ship was completely overturned.'[2]

Instead, when the ship began to tilt to starboard, Yu had ordered a hard a-starboard turn so it would be back facing into the wind. This had increased the number of fatalities.

The tribunal offered a 'Lessons Learned' chapter. Much of the content was basic seamanship. All vessels sailing across the Antarctic Ocean should be braced for bad weather. As the seas were given to frequent and dramatic changes in weather and sailing conditions, doors and openings should remain closed. Safety information should be written in the languages that crew use.

Noting that hiring agents had been involved, the tribunal added: 'The implication of this hiring policy was that the ship owner did not even have full knowledge of the names and nationalities of the crew members on board the sunken vessel.'

Only five bodies were recovered. When I tried to put names to the dead, Insung Corporation ignored my request for information. A week after the sinking, the corporation website was still reporting that the vessel was fishing, and proclaiming: 'Our vision is to integrate the different areas of fishing to provide consumers with fresh fish at lower costs faster than any other company.'[3]

Tearful family members of the missing sailors gathered at Insung's

Seoul office and waited for news. Kim Sun-su, a brother-in-law of Yu, hoped the missing skipper would return home alive. Yu had told him in a recent phone call that he didn't want to work on the trawler any longer.

When Insung provided Korean names to the Seoul media, it was revealed that the dead included first mate Choi Eui-jong, 33, chief engineer An Bo-seok, 53, and first engineer Ha Jong-geun, 48. Cook Jo Gyeong-yeol, 55, and crewman Kim Jin-hwan, age unknown, were missing.

Four of the dead were Vietnamese. *VietNamNet*, a web arm of the state-controlled media, ran a story under the headline, 'Thousands of expatriate sailors face risks and difficulties'. The public, it reported, was 'paying special attention to the sunken Korean fishing vessel in the Antarctic'.[4]

Nguyen Thi Ngan grieved for her son Nguyen Van Son. 'Why is it so unfair? A poor one who has to work away from home is killed.'

Nguyen Tuan and his wife Dang Thi Lan were disconsolate. 'Oh my poor son! We've lost you now,' the wife sobbed. After finishing high school, Nguyen Tuong, who was the couple's eldest son, had taken the fishing job so his family could escape dire poverty. He had died just three months after the family took out loans worth US$769 to pay his way on to the ship. Tuong's uncle, Nguyen Song Hao, was also among the crew members listed as missing, presumed dead.

'Cries of grief resounded in the two small communes of Kỳ Khang and Kỳ Ninh in Ha Tinh Province's Kỳ Anh District as relatives mourned the deaths of their relatives,' *VietNamNet* reported.[5] A district official said that in recent years many locals had left to work abroad in the hope of escaping poverty. 'More than 500 residents … of whom 100 are working on Korean ships. Eight of them have died, mostly in Taiwan and Malaysia. … Many have had to borrow money to pay the fees for getting the jobs.'

When a reporter visited the home of one of the missing men, Nguyen Van Son, his mother and wife were in mourning. Son had been married two years and had gone to South Korea to work and make money to support his family. A neighbour added, 'He went to work when his wife was pregnant. Now his one-year-old son has lost his father.'

Not far from Son's house, Nguyen Van Thanh's parents were distraught. For a time they had held out hope for survival 'but today I was told that he was missing and the rescuers said that he is dead. It's very painful,' the father said.

It was reported that more than 1,000 Vietnamese sailors worked on Korean fishing ships and a similar number on Taiwanese vessels. Young sailors earned US$180 a month and experienced sailors $210.[6]

The price of toothfish continued to rise, and nobody seemed to take much notice of the shocking state of the Southern Ocean fishing fleet. Just a year later, in December 2011, a Russian-flagged, California-owned, 23-year-old boat called *Sparta* hit an iceberg in the Ross Sea while hunting toothfish and nearly sank. It took a multinational effort, including two parachute drops from the Royal New Zealand Air Force, before the ship was able to get out of the ice. No one was hurt. A month later, in January 2012, a Korean boat, *Jung Woo 2*, caught fire in the Ross Sea, killing three of its 40 crewmen. Its sister ship *Jung Woo 3* came to the rescue before the ship sank. Both ships had been with *Insung 1* when it dropped to the bottom of the ocean two years earlier.

New Zealand Green Party member of parliament Gareth Hughes was one of those who wondered why the fishery was operating. 'In little over a year, three fishing vessels – the *Jung Woo 2*, along with *Sparta* and [*Insung 1*], which sank with 22 lives lost – have come to grief in the Ross Sea. … This pristine environment must not be put at risk by old, single-hulled unsuitable fishing boats like these, [which] race to catch as much as they can despite the weather…'[7]

After the rescue, *Jung Woo 3* sailed on to Montevideo. There it caught fire, killing two Vietnamese crewmen and injuring six. Next, its sister ship *Jung Woo 1* had a fire in the same port. This one did not cause any injuries or serious damage, but two Vietnamese men were arrested for causing it.

Sparta would return to my watch list later.

3 Crime

Thanks to Adam Smith's 'invisible hand' in a global marketplace, consumers mostly live happily with anopsia. We do not know, and do not want to know, for example, that the fish in countless processed fast foods will have been caught by men and boys working in slave-like conditions, or that women working in sweatshops around Asia will have processed it. If it has been caught in New Zealand's exclusive economic zone, it will be entitled to call itself 'Produce of New Zealand' and no one will be any the wiser. It is like the days of William Wilberforce: who was to know that the sugar in your cup of tea had been produced by slaves?

These days plenty of buzzwords are bandied around in the media. Human trafficking and slavery are common examples. Being rather cynical, until I found myself covering the fishery beat I had not given a lot of thought to either. Now I had learned that much large-scale fishing involves taking men from poor countries and putting them into an internationalised environment, where they are largely beyond the reach of police and regulatory authorities, and in a realm of vague and difficult-to-enforce treaties and multinational rules.

As foreign charter vessels became increasingly involved in New Zealand's fishing industry, neither politicians nor government officials seemed to me to take the slightest interest. The men on board the ships were invisible; the politicians and officials may not have been paid not to see but they knew enough not to look.

Many countries face the temptation to allow fishing in their exclusive economic zones to be contracted out in licensing systems that survive on migrant labour contracts. If there is a wilful blindness by authorities to the human exploitation involved, it is because they believe that without it a fishery will be 'uneconomic'. A New Zealand government inquiry in 2012 heard what 'economic' means. A local factory trawler has an average annual wage bill of around $5.6 million; cheap Asian boats run an annual wage bill of around $676,000, and most of that goes to the officers.

This has made marine living resources a high-profit, low-risk target not only for ruthless companies but also for criminals. Coming to terms with the linkage between fishing and a variety of criminal activities is key to understanding the wider issues, from abuse of workers to despoliation of the oceans.

That fishing and crime go hand in hand may not seem obvious to the millions of people who buy their fish frozen, packaged or crumbed, in a branded box, and ready for a quick fry-up in a pan. After all, the branding usually involves recognised and trusted companies, which are presumed to take a careful approach to protecting their supply chains.

This, though, can be an illusion, as the 2011–12 European scandal around the presence of horsemeat in processed food showed. In Britain, five major retailers, including Tescos, were found to be selling beef burger products that contained significant amounts of horse meat.

But fish is fish, you may say, and surely buying whole fish at a market, where there does not seem to be much distance between the fisher and the consumer, removes the opportunity for criminal activity?

Unfortunately not. As global fish stocks decline, the surviving stock becomes worth more. Access to commercial fish stocks is usually government-controlled, but in parts of the world, because of corrupt officials, quota can be illegally bought or catch limits ignored. In covering the South Pacific I have regularly come across fishing quotas – usually for tuna – being held by friends, families and allies of governments in power. One vivid example was in the Solomon Islands under the premiership of Solomon Mamaloni. In Mamaloni's last term in office from 1994 to 1997 his personal control over fisheries access enriched him significantly.

A raft of studies have also found fish fraud to be big business. Between 2010 and 2012, researchers from Oceana, a Washington-based lobby group, collected over 1,200 samples of fish from nearly 700 US retailers and DNA-tested them. A third were not what they were labelled as. The most mislabelled were snapper, 87 percent, and tuna, 59 percent. Nearly half of all retailers, restaurants and sushi shops were selling mislabelled fish.

Fish fraud is not new. In 1989 the US Food and Drug Administration confiscated 20 tonnes of oreo dory that had been imported from New Zealand as expensive orange roughy. Orange roughy sells for about three times the price of oreo dory.[1]

The other criminal activity comes about because the global marketplace constantly requires margins to be squeezed so food can be produced cheaply and sold to customers at prices that will knock out the competition. If a supplier can make men and women work for a pittance, breaking laws in the process if need be, margins can be trimmed and the customer at a supermarket in Auckland or Sydney can get the product cheaply.

In 2011 the United Nations Office on Drugs and Crime published 'Transnational organized crime in the fishing industry', a paper based on an extensive search of government, media and international political and scientific studies. 'Perhaps the most disturbing finding … was the severity of the abuse of fishers trafficked for the purpose of forced labour on board fishing vessels', the agency reported.[2] Practices uncovered were cruel and inhumane in the extreme, with men held as virtual prisoners. Deaths, severe physical and sexual abuse, coercion and a disregard for safety and working conditions were common. Around 22,000 people from Laos alone had been taken against their will into the Southeast Asian fishing industry.

The Environmental Justice Foundation, a London-based body that monitors abuses in the fishing industry, had highlighted the case of an illiterate Nepalese fisherman who signed up to a Singaporean contract using his thumbprint, and agreed to a three-year contract with a monthly salary of US$200. A Singaporean agency was to retain 75 percent of the money; payment for the first six months of the contract was to be made only when the full three-year contract had been completed. Repatriation costs were not covered, and the crew member would be abandoned in the nearest port if the contract were breached. A breach was defined as 'own sickness, lazy and rejected by captain etc'.[3]

Working hours were around 18 hours a day ('sometime more, sometime less') and there was no overtime pay. Some food would be provided but noodles and biscuits had to be purchased by the crew member himself. 'Sea water will be used for bathing and laundry purpose.'

Should the crew member be unhappy with the arrangement, the contract stated that he had 'fully understood' that 'I will not claim back any amount of money I spent for securing this job'– that is, the recruitment fee. He was expected to 'work hard, obediently and diligently' about 19,710 hours over a three-year period for an average pay of US37 cents an hour. From this, the cost of the recruitment fee and repatriation – 'minimum US$2,000' – had to be deducted, making the crew member a potential victim of debt bondage.

In many of the ships the foundation had investigated living conditions were abysmal. Sleeping quarters were often cramped; there had been reports of shared bunks with cardboard mattresses, stacked less than a metre above one another. Cooking facilities might be unhygienic and food supplies limited. Men had become severely malnourished and fallen ill through excessive exposure to sun and sea water.

To add to the catalogue of horrors, the United Nations body noted that the victims of 'trafficking for the purpose of forced labour on board fishing vessels' were often subjected to severe physical and psychological abuse. 'Disobedience or lack of effort is often struck down forcefully. There are a number of reports of physical injuries and deaths induced by senior crew. There are sources that suggest that victims at sea have been tossed overboard when sick, injured or dead. Fishers that fall overboard are sometimes not rescued.[4] There had been news reports of bloated corpses of foreign fishers washing up on coasts.

The International Transport Workers' Federation had reported that a Filipino fisherman aboard a fishing boat engaged in illegal fishing, known in the legal world as 'illegal, unreported or unregulated fishing' – IUU – had been chained up for 30 days in a two-square-metre storeroom and beaten with a baseball bat.

'We often had to sleep with our work clothes and sometimes wet working clothes,' the man had testified. 'We were denied medical treatment and medicine. … only permitted to eat what was left after the … crew had eaten and were left with half-finished cups of coffee to drink and food left over. We were required to massage … officers and crew on a daily basis after our long hours of work. We were punched, kicked and beaten on the head with closed fists by the … personnel regularly. The crew often grabbed our sensitive parts, applied pressure to the extent that we cry in pain. They also squeezed our necks until we fall to our knees.'[5]

The nationality of the vessel concerned was withheld, but the man's experience sounded much like a case the federation had found on a boat fishing in New Zealand's exclusive economic zone.[6]

Slave labour has become a common feature of fishing worldwide. As one victim summed it up: 'We were always thinking of escaping … there was no way, though. We were powerless. The sea itself was our prison.'

Much of this has been documented by the United Nations Inter-Agency Project on Human Trafficking. The agency has reported, for example, that thousands of Cambodians are trafficked each year. 'Some of the worst exploited are the men and boys who are deceived on to long-haul fishing boats that fish the waters of the South China Sea, including into Malaysian waters.'

These boats go to sea for up to two years at a time, during which crew members endure inhumane conditions. In a report on 49 Cambodian men and boys who had been trafficked on to Thai long-haul fishing boats, the project found a common theme – 'deception and debt bondage by two or more Khmer and Thai brokers; the men's sale to a Thai boat owner for 10,000 to 15,000 baht [US$300 to $450]; slave-like working conditions at sea, including beatings, deprivation of food, inhumane work hours (for example, working three days and nights straight when nets need to be mended); lack of medical treatment for illnesses and injuries; threats of death; and sometimes, reportedly, murder.'[7]

Those who choose to jump ship in Sarawak, a Malaysian state on the island of Borneo, fare no better. Many end up in a new kind of slave trade, forced to work on palm oil and rubber plantations. Others are picked up by Malaysian police, who cane and deport them – or sometimes hand them back to the ship they sought to escape. Nineteen-year-old Banteay Meanchey told the UN body that on his boat crew members were beaten frequently. And there was worse. 'On shore [at Sarawak] we saw a Thai captain decapitate a Vietnamese fisherman, and another Thai captain decapitate a Thai fisherman.'[8]

Another recruit, Prey Veng, said he was abused, beaten, forced to do hard work, and even threatened with being shot. 'There were a lot of big waves. It was not safe. A worker died because he fell into the sea and my boss knew it as well, but he did not return the boat to save him.'

The twenty-one year old was supposed to get 6,000 baht a month, but each time he asked he was told to wait. After a year he was given only 20,000 baht. 'I sent all of my money to my mother in Cambodia through [the broker]. But [the broker] did not send my money to my mother. He kept it for himself.'

What I was learning first-hand about abuse in the fishing industry involved men, but the United Nations Office on Drugs and Crime reported that children were also victims. In Vietnam's polluted Mekong River, for example, shrimp and catfish – marketed as basa – are caught using child labour. In West African states such as Senegal, investigators have found children being used on purse seiners to herd tuna. 'In purse-seining operations,' the office reported, 'children are required to dive into the water to guide the fish into the net, and this is generally done by those aged between 12 and 13 who have some swimming experience.'[9] A high risk of drowning is normal because of the long time the children spend in the water, far away from the adults, who are busy dealing with the nets.

In vessels that go to sea for days, children are used to bait hooks, fish with hand lines, and haul in longlines. They also arrange the catch in iceboxes, help prepare meals, and face the risk of physical abuse.

For some activists, these issues around the abuse of both adults and children in the fishing industry can seem a distraction from other issues, such as sustainability of fishing, conservation of fisheries, and economic implications. But as far as Steve Trent of the Environmental Justice Foundation is concerned, competition about which issues are the gravest is irrelevant when 'we have detailed evidence on film of pirate vessels involved in human trafficking, the arms trade, narcotics and even murder'.

In 2010 the foundation found Senegalese boys as young as 14 on *Marcia 707*, a South Korean-flagged vessel in Sierra Leone waters. When officers boarded the ship they unearthed a makeshift structure used to house up to 200 people, including the boys, in cramped, unsanitary conditions. The boys said they had been picked up by the ship in Senegal and were being forced to work for three months at a time.

Living conditions aboard another Korean-flagged boat, *Apsari 3*, were even worse. The 36-man crew from China, Vietnam, Indonesia and Sierra Leone described being flown from their native countries to Las Palmas in the Canary Islands. Their contracts were set for two years, with no chance of a visit home. One man had never seen his 18-month-old child.

For sleeping quarters, eight men shared a small area of the hold that contained four bunks made from planks and cardboard. They took it in turns to sleep in the windowless space, which led directly into the fish hold, while the others worked their long shift. Crew from Sierra Leone on board the boat had no contracts, and were paid in boxes of frozen discarded 'trash' fish, which they had to sell in local ports.

Horror associated with fishing is ubiquitous. A US State Department report revealed that in August 2006 more than 30 Burmese fishermen died from infectious diseases and lack of medical care on fishing vessels off the coast of Thailand. The bodies of the victims had been tossed overboard, 'discarded like common refuse'.[10]

IUU – illegal, unreported or unregulated – fishing takes place over the horizon, largely away from public view. As well as human degradation, it involves what may at first seem like only technical crimes: fishing more than the catch limits; using outlawed gear; 'trucking', in which a ship catches a species of fish in one area and then goes to another area and later claims it caught the fish there. In this way, fishing companies can hide the fact they have taken fish from a vulnerable area. In another IUU action, 'transhipping', two ships rendezvous, usually on the high seas, and transfer cargo to obscure how much fish has been caught and where.

The cumulative effect of these activities can be a severe reduction in fish stock in the fisheries targeted, and even the collapse of vulnerable fisheries. In 2013, Oceana listed some of the stolen seafood. In the case of species such as Russian sockeye salmon, 60 to 90 percent more is caught than is officially reported. Illegal catches of toothfish are five to ten times greater than those officially recorded. Half of all swordfish and tuna catches in Greece and cod catches in Britain are illegal.[11]

Oceana believes sales of Pacific bluefin tuna on the black market may bring in US$4 billion annually, and that the volume of this species illegally caught is five to ten times higher than the catch officially recorded. Illegal catches of other tuna – skipjack, yellowfin, albacore and bigeye – are worth around US$548 million a year.

Fishing companies, vessel owners and even governments are often actively involved, as is organised crime: the UN and other agencies report that high-value, low-volume seafood is a currency, forming part of the trade in drugs and arms as well as human trafficking.

Those involved are clever: the vessels they use to steal fish are old and unsafe. If they are apprehended by authorities and the boats forfeited, they suffer no great economic loss. In addition, the vessels are often registered to nations that are unable or unwilling to prosecute, even if they could: much modern ship-owning is characterised by complex arrangements involving front companies and layers of shell companies across several borders.

Namibia – the former German colony of South-West Africa that achieved independence in 1990 – knows a thing or two about fishing crime: its government agencies have in the past provided flag cover for illegal fishing operations. Mozambique, on the opposite coast, is just as implicated: its operations have included giving credibility and documentation to IUU-caught fish.

These countries, which came into existence through liberation wars in the lifetimes of many of their politicians, have, however, begun to realise that short-term gain is a recipe for international disaster. In 2007 Namibia launched a national plan of action against IUU fishing.[12] In 2008 the Namibian minister of the fisheries and marine resources, Abraham Iyambo, warned that his was the last generation that could end 'the troubling destruction of our oceans' and the hardships it brought to people on many coasts. 'It is not an exaggeration to state that the plague of illegal fishing is one of the largest environmental crimes of our time,' he said.[13] Mozambique has also taken steps to avoid having fish documentation accepted at face value.[14]

Abalone – or pāua in its New Zealand form – is heavily linked to organised crime. The Chinese appetite for this shellfish makes it an ideal currency for transnational criminal groups trading in drugs and in precursor chemicals for the manufacture of amphetamine-type stimulants such as methamphetamine.

The New Zealand Police's Organised and Financial Crime Agency estimates that about 39,600 tonnes of pāua, both legally and illegally harvested, reaches the international market each year. The amount varies according the extent of enforcement, and demand outstrips supply in Asian countries, where there is a belief that the shellfish delays senility and increases fertility. Much of this pāua is farmed, but each year about 4,400 tonnes is poached. The agency has reported that both local gangs and Asian-organised crime groups are involved in this poaching, and also in the poaching of another high-value commodity, rock lobster. One study found that in 2007 a black-market pāua merchant with links to Asian organised crime was able to buy illegal pāua for NZ$17 a kilogram and on-sell it for NZ$46. Moving 1.6 tonnes each two-month period, he made a yearly profit of NZ$270,000.

New Zealand abalone is regarded as inferior to Australian and Japanese varieties, but according to the Ministry of Justice pāua smuggling is a significant area for organised criminality. In operations similar to those seen in Australia and South Africa, divers are hired on an individual basis, or as part of a gang. The diving takes place at night and the pāua goes to Asian-organised crime groups via middlemen.

Operations by the then Ministry of Fisheries in 2001–02 (codename 'Pacman') and 2008 ('Paid') led to around a hundred people being charged with serious crimes. Operation Pacman found that bags of pāua were being added to the luggage of members of large tour groups. More recently, pāua has been found leaving the country in tins bearing false labels such as 'Milk Powder'. Operation Paid discovered pāua was being paid for with illegal drugs.

In 2011, Operation Fusion, an Australian Federal Police crackdown on abalone smuggling, reported an 'intricate system of dive locations, abalone theft, surveillance techniques, storage locations, transport operations and illegal trade in Sydney'.[15] Police seized 380 kilograms of abalone worth AU$60,000.

In 2004, another shocking connection between crime and fishing was dramatically revealed when 23 Chinese cockle pickers were caught by the tide in Morecambe Bay, England, and drowned. Investigation showed that a snakehead gang of people smugglers had been paid US$31,000 a head to get the men and women into England. The cockle pickers, who earned US$4 an hour, were expected to pay back the debt. The man behind the crime, 29-year-old Lin Liang Ren, was an accountant from a wealthy family in China. He had faked studies in London while running the gang of illegal immigrants, who were sent out to scour beaches for cockles.

After the tragedy Lin tried to slip away, claiming he was just an ordinary worker. Under arrest, he warned survivors that there would be 'harmful consequences' for their loved ones at home if they cooperated with police. The UK Press Association reported that he was true to his word: there were assaults and threats in China. In 2006 Lin was given 14 years' jail; in the words of the judge, he had cynically and callously exploited his own people.[16]

Fishing boats themselves play a role in many crimes, from drug smuggling, people smuggling and weapons trading to acts of terrorism. The November 2008 Mumbai attacks that killed 166 people over four days began with the hijacking of a fishing trawler. Ten terrorists forced the *Kuber*'s captain to sail to a spot a few kilometres off Mumbai, where they killed him, boarded three inflatables and travelled to the shore. It says something of the status of fishermen that when several saw strange men coming ashore in inflatables and called the police, the police did nothing.

Fishing boats are favoured by criminals because of their global reach and their fish distribution networks. Some fishing companies use complex incorporation and vessel registration strategies, together with logistical coordination of vessel support services at sea, to avoid detection. These days there are transparent and easily accessible internet systems – available even on smartphones – for tracking merchant vessels. The same is not true of fishing boats. While they can be tracked in port, at sea their positions are often cloaked in secrecy. This goes back to a belief that wild fish are a property right, and that if a skipper finds a rich ground he should be able to keep it secret. All legal fishing boats today are supposed to carry position-indicating systems – known variously as VMS (vessel monitoring system) or AIS (automatic identification system) – which provide authorities with catch records but, as part of the deal, the authorities assiduously keep the information secret from the ships' competitors and the media.

When a Pacific Islands Forum Fisheries Agency came into existence in 1979, it demanded that all boats catching tuna inside the exclusive economic zones of its 17 member states be equipped with vessel monitoring systems. In theory, this would end the need for surveillance and enforcement: with a click of a mouse, authorities would be able to see the position of a fishing boat, the size of its catch, and the species caught and where. Resistance from Asian fishing countries, including Japan, was strong. Each year, when Pacific leaders held summits around the region, the secret, and therefore unreported, battle was the attempt to get VMS installed.

Alongside this was another battle over which of the many VMS systems to use, with some Pacific politicians taking bribes from companies and nations anxious to push their hardware. Even this was not necessarily the biggest problem for the Pacific fishery – many Pacific states have long operated against a background of corruption and inefficiency. But once VMS was installed on boats, there was an issue of what to do with the information.

Early in December 2012, the Western and Central Pacific Fisheries Commission, a body set up seven years earlier to conserve and manage stocks of tuna and other highly migratory fish, met in Manila to divide up tuna stocks in the region. By this time, there were estimated to be over 2,000 vessels on VMS in the Pacific, each reporting every four hours. However, buried among the commission's data was the revelation that much VMS data was not being acted on.

One instance concerned a place known as Eastern High Seas Pocket, a kidney-shaped blank on the map between the Cook Islands and French Polynesia. The Cook Islands had come up with data showing that the pocket was subject to extensive illegal fishing, and was also an access point for fishing boats wanting to illegally enter nearby exclusive economic zones. VMS was showing the offending boats, and the data was being electronically recorded, but no one was especially monitoring the data and no action was being taken. Although VMS remains a sound concept, for many economically pressed Pacific states, it's just one more thing they need to do on next to no money.

Somalia offers an unusual example of the extreme connection there can be between fishing and crime. The country's breakup and descent into civil war presented an opportunity for large-scale Asian fishing companies, ignoring any environmental considerations, to strip the country's waters. Somalians, who fished by hand off small boats, witnessed large factory trawlers sweeping in and vacuuming the area of fish.

This criminal activity led to local people resorting to another kind of crime – the hijacking of ships for ransom money. In time, this became more lucrative than fishing. Many Somalians with small boats saw little point in returning to fishing even after the Asian fishing boats, which had been their early targets, left fearing for their safety. 'I used to get less than $200 a month from fishing but now that poverty is gone,' a Somali-based pirate – who gave his name only as Gurey – told Agence France-Presse. 'You never know when this business is going to be over, but what I tell you is that going to the ocean for piracy is better than going out for fishing.'[17]

In a strange twist of fate, the pirates' actions, although criminal, have turned out to be good for the environment. Even before the civil war, the Indian Ocean's large maritime resources had for decades attracted industrial fishing fleets from Europe and Asia, which went on month-long campaigns to feed their demanding domestic markets. While the Somalian pirates became the curse of the seaways, they also deterred many foreign fishing boats from plundering the ocean.

In 2010 Agence France-Presse reported that there had been bumper fishing since the pirates were active. 'This is a good season,' said Aziz Suleiman, who co-owned the little wooden shack where he and his partners auctioned off fish. Nineteen-year-old fisherman Zedi Omar agreed. 'We no longer see the foreign fishing boats we used to see. We only spot cargo ships.'[18] Spanish, French, Taiwanese and other, mainly tuna-fishing, trawlers based in the region have to delay their campaigns in order to hire security, and use elaborate new routes to dodge the ransom-hunting sea bandits.

What this has demonstrated, rather ironically, is that ending the commons system of the ocean could be the solution to overfishing and environmental degradation. Place open oceans under some kind of regulation, such as vastly expanded exclusive economic zones, and a degree of control could finally be exercised.

A British-based group Havocscope, which monitors the world's black markets, reported that in 2012 illegal fishing was worth US$23.5 billion, comparable to international methamphetamine trading and human trafficking.[19] Illegal taking of tuna in the Pacific alone is said to be worth US$1 billion.

But as I was to discover, illegal fishing involves more than just shadowy characters with organised crime connections. Some big, mainstream, high-profile fishing companies deliberately, or accidentally, run pirate fishing boats. Sometimes this is based on little more than bad paperwork; at other times it is a contrived criminal attempt to maximise profits.

Havocscope has compiled startling data. For example, between 2001 and 2011, Chinese boats caught 3.1 million tonnes of fish off the coast of Africa and 80 percent of it was illegal. The value of this illegal catch was put at US$12 billion. And this is not just a case of stealing fish stocks and avoiding licence payments to the local governments: in many affected areas local fish is a vital part of the diet, and if fish stocks diminish the impact is disproportionate.

It is not just poor countries that are hit. Each year, according to Havocscope, 27,800 jobs are lost in the European Union as a result of losses due to unregulated fishing. Illegally caught fish make up 16 percent of the European Union's yearly catch.

In a graphic piece on illegal fishing, English journalist David Smith reported that the problem was now on a 'colossal scale'.[20] In Africa, where the losses affect some of the world's poorest coastal communities, locals are fighting back. In Sierra Leone, fishermen from 23 fishing communities, backed by the Environmental Justice Foundation, secretly filmed ten international pirate trawlers and sent in 252 reports to the authorities. 'If these communities don't have fish they go hungry and possibly even starve,' Steve Trent said. 'In Sierra Leone, fish represents nearly two-thirds of animal protein.'[21]

This project, in which fishermen are given basic camera equipment or smartphones by the foundation and instructed to take pictures from their dugout canoes, costs just US$200,000 a year. It has already caused a 90 percent decline in illegal fishing in the country's territorial waters within ten miles of the shore over a two-year period. In the next three to five years, as the project is extended to Liberia and Ghana, it promises to have a massive impact in West Africa.

The foundation knows that, without a curb on illegal fishing, the world faces serious collapses in whole populations of fish and ecosystems. 'Pirate fishing exacerbates the problem of overfishing of 37 percent of the world's fish stocks,' Trent said. 'They use illegal gear, such as monofilament nets, that have a very small mesh gauge and scoop up everything in their path. Then they throw around 70 percent of the fish over the side of the boat, which is crazy in our resource-scarce world.'

The foundation has witnessed the human effects of its campaign. 'Again and again and again people on the beaches tell us, "This is amazing; our fish are coming back and we're getting bigger and bigger catches since we got rid of the pirate trawlers." The individual fish sizes are also growing, which means the populations are recovering.'

4 Oyang

On March 9, 2010, Hyonki Shin sailed South Korean-flagged *Oyang 70* out from Port Chalmers in Dunedin, New Zealand. The 42-year-old had recently taken command of the 38-year-old ship; his predecessor had died after drunkenly falling into the harbour.

Even by the dubious standards of deep-sea fishing, *Oyang 70* was ready for the knacker's yard. The 1,698-gross-tonne, 74-metre-long stern trawler had a reputation among harbour pilots. They called her 'tender', a polite way of saying her stability was marginal and her helmsmen needed to be alert.

That *Oyang 70* looked like a rust bucket to the untrained eye didn't mean it was beyond redemption: old fishing boats can remain useful as long as there are warnings about safety. Plainly, Maritime New Zealand inspector Peter Dryden felt something along those lines when he inspected *Oyang 70* before it left. On his report he wrote: 'Vessel floats.'

Later he would come to regret this. 'With hindsight,' he said, 'I can see that it would be possible to misinterpret my words.'[1] The reality was that owners of foreign charter vessels such as *Oyang 70* were indifferent to the real issues of safety; as long as a boat passed a legal mark they gave it no more thought. Dryden had reported what he was required to report, and what he saw was a floating ship.

Oyang 70 was part of a global fleet owned by the Sajo Oyang Corporation. One of South Korea's largest fishing companies, Sajo Oyang was keen to establish global dominance in fisheries. The ship's crew of mostly Indonesian and Chinese men, plus one Filipino, were aware their ship was not terribly seaworthy, but they had no power. They were essentially sweatshop labourers, ensnared in the Korean fishing system.

As I spent more time covering the fishing industry, I would find the brutality of the Korean companies a mystery. As a nation, South Korea goes out of its way to ensure the world sees it as sophisticated and cutting edge, a place with high broadband penetration, smart cars and clever people. And yet, as hunter-gatherers on the world's oceans, its fishing companies choose to be primitive and uncaring.

Also aboard *Oyang 70* that day was Serge Artieu, a New Zealand Ministry of Fisheries' observer. Artieu's role was to ensure that fisheries rules were followed. The boat could catch only so many fish; the fish had to be of a certain size; and they had to be caught according to the rules for each species. There was also the politically sensitive matter of bycatch – fish species or other wildlife caught accidentally. Boats fishing in New Zealand waters often snaffle the food of birds such as albatrosses and marine animals such as seals, but this becomes an issue to authorities only if the affected wildlife are directly harmed.

Oyang 70 was heading south to catch Auckland Island arrow squid. The subantarctic islands, collectively designated a UNESCO World Heritage Site, are home to the endangered New Zealand sea lion, one of the world's rarest, which lives on squid and is attracted to boats hauling in big loads. In any given season, the Ministry for Primary Industries sets an acceptable number of sea lions that can be killed. If the number is exceeded the fishery can be closed. Fishing boats are required to use a sled that works to keep sea lions out of nets.

Boats also have to be rigged with exclusion devices – wires that hang out on both sides of the stern – to prevent birds going into the nets. Artieu reported that on the voyage to the Auckland Islands the boat caught a white-capped albatross and three white-chinned petrels, killing two of them.

The main fish bycatch was spotted dogfish, or rig, a small shark species that, when legally caught, is often destined for fish and chip shops in New Zealand. *Oyang 70* recorded its bycatch, because it was required to, and then dumped it. Many trawlers have facilities to process waste and fish bycatch into fishmeal. The product is not worth much but has become a useful feedstock for Chinese eel farmers. *Oyang 70* had no such facilities, and its dumping was illegal: the law requires the catch to be landed and, if it is a quota species, the deemed value paid.

On its first trip with Hyonki Shin as skipper, *Oyang 70* ran 53 tows, each around 13 hours, adding up to about 28 days. It filled all its storage capacity with squid, winning it away from the sea lions and the equally endangered yellow-eyed penguins. Most went to Japan, the world's biggest consumer of squid, or to China.

Artieu's reporting form required him to deal with some aspects of the ship itself, but not to observe the crew's conditions, or whether the officers were following New Zealand labour and human rights laws. These laws, supervised by the Department of Labour, do not concern themselves with the welfare of fishing crews on foreign charter vessels. The only reason observers have to report on some aspects of conditions onboard is because the department has health and safety responsibilities towards its employees – the observers themselves.

The department's concern for its observers is not misplaced. In one infamous case, which came to public notice only when I asked a question under New Zealand's Official Information Act, an observer had ended up in hospital having bedbugs surgically removed from his back after a tour of duty. Other observers had contracted tuberculosis. And in January 2014 a 47-year-old observer on a Korean boat, *Sur Este 700*, had to have his forearm amputated after his life jacket became snagged on a conveyor belt as he was putting it on.

Sajo Oyang had created a New Zealand-registered company, Southern Storm Fishing (2007) Ltd. Although controlled from Seoul, Southern Storm looked like a New Zealand company. Oyang was not alone in creating online shell companies in New Zealand – such a company can be created from anywhere in the world for just NZ$163.55. These entities are set up for numerous reasons, ranging from tax avoidance, to hiding the true nature of company ownership, to being fronts for illegal activities such as money laundering.

Southern Storm had contracted with another company, Fisheries Consultancy (NZ) Ltd of Lyttelton, Christchurch, to manage the ship's port needs. Before the voyage, Fisheries Consultancy's Russ Barron took Artieu on to *Oyang 70*. When the issue of safety came up, Barron said that Artieu should 'stick by the captain at all times in event of real fire'. Sticking by Hyonki Shin was never going to be a good option: Artieu noted that the motor of the ship's one rescue boat, a Zodiac, was not working.

Much else was dodgy. The water was of 'dubious quality, often rusty, certainly not drinkable,' Artieu noted. The fact the fresh water was rusty suggested *Oyang 70* might be having seaworthiness issues.

'Hygiene on the vessel needs some improvement, especially in crew living area,' Artieu also found. 'Crew toilet has no running water to wash hands, nor paper towels. … The bath has no shower, only salt and freshwater plunge pools heated by steam.'

As for his own conditions, Artieu ruled them satisfactory: he had a ten-square-metre cabin. However, once back on shore he would complain that, while he had found a 'small engineer's room' in which to keep his clothing, it had been cramped with 'rusty unused motors and other old parts'. More seriously, the factory deck had been difficult and dangerous to move about on and he had kept hitting the ceiling.[2]

As Artieu settled into his cabin the ship's cook gave him three dozen cans of beer. It would later be revealed that this was an Oyang strategy for nobbling observers so illicit activity could be carried out at night. An observer, if not drunk, would at least be sufficiently relaxed to go to sleep. Informants would tell New Zealand government investigators that, once an observer was asleep, 'high-grading' would take place. This process involves dumping inferior fish overboard in the expectation that the boat will get better quality catches later. It is illegal.

Artieu reported that on three occasions he saw plastic bags and rubber gloves in the water near the ship. These may have come from the ship as the crew covertly unwrapped and dumped fish. There were no other vessels in the area.

After returning to port, *Oyang 70* went out again in July. When it arrived back at Port Chalmers at 7.35 a.m. on August 12, 2010, its catch was quickly discharged into containers destined for South Korea; although untouched by any New Zealand hand, it would be sold as 'Produce of New Zealand'.

5 Southern blue whiting

Oyang 70 was scheduled to leave port again the afternoon of her arrival. A pilot came aboard and the tugs were made fast alongside, but the ship's main engine would not start. It was not until ten the next morning that the ship was finally able to sail. Even as it left the harbour, the vessel was 'tender', or of doubtful stability, the port pilot later recalled.

Oyang 70 should have had on its bridge a stability book, a manual giving formulas for balancing the ship in various circumstances. I was aware of the importance of such a book, and of having officers who understood its instructions. In 2009 I had covered the tragedy of *Princess Ashika*, a 37-year-old ferry that sank in Tonga with the loss of 74 people. There had been a stability book on board *Princess Ashika*, but it had emerged that those who commanded her voyages did not know what it was for.

Vessel stability can be intuitive to an experienced band of officers but no skilled seaman would rely on instinct alone. Like the problems in high school logarithm books, stability involves a number of different calculations. And each voyage is different. A careful skipper will check fuel and cargo weights, their positioning in the ship, and the weather.

Hyonki Shin was not such a skipper. After the July voyage, the electrician on board *Oyang 70* had described the ship 'heading into waves

… water inside the ship was shifting from side to side and due to the weight the rocking motion was slow'.[1] He was 'very scared, as were the rest of the crew' but they did not go near the captain. He was 'an angry man'. They feared they would lose their jobs.

Shin headed south-southeast toward the 13 islets and rocks that make up the Bounty Islands, which lie 670 kilometres east of the southern city of Dunedin. These islands got their name in 1788 when the HMS *Bounty*'s captain, William Bligh, heading towards a mutiny, came across them. In the nineteenth century the islands were favoured by sealers, who took as many seals as they could until there were no more. Shin's target was the southern blue whiting, which in August and September aggregate to spawn at between 250 and 500 metres where the Campbell Plateau descends into the Bounty Trough.

Southern blue whiting are a low-value, high-volume catch, worth around $1.50 a kilogram aboard the boat. Orange roughy, by comparison, are worth $17.36 a kilogram. Whiting are worth catching only if the costs involved are pared to the bone. The way to do this was with a crew working for almost no wages, and on a boat with heavy subsidisation. Sajo Oyang Corporation was entitled to heavy tax breaks in Korea, and its catch, even if made in the name of a New Zealand company that had chartered the boat, was tariff-free.

Southern blue whiting is not a traditional catch. It is only as a result of recent science and technology that the stock has been discovered and uses for it invented. It is a factory fish that can be filleted easily and can be fried, baked, microwaved or steamed. Its mild flavour is regarded as ideal: it neither particularly smells like fish nor tastes like it.

Worth around US$19 million a year, it is often turned into surimi, surimi being the Japanese word for ground fishmeal, a key ingredient in fake crab sticks, fish sticks and pretend lobster. A lot also ends up in pet food – fishermen risk their lives in the effort to feed the world's cats.

The southern blue whiting aggregates heavily in spawning, and sonar-equipped ships such as *Oyang 70* can easily locate plumes of between 1,000 and 2,000 tonnes. A catch of between 100 and 150 tonnes is possible in minutes.

The fish, though, are oily. When hauled out of the water they tend to become a fluid mass on the deck, creating what is known as the free surface effect, a kind of movement that can endanger a ship's stability.

Oyang 70 was to catch this southern blue whiting legally. The ship's owners had purchased rights to it in a complex New Zealand trading system in which small, mainly Māori-owned blocks of quota are consolidated and onsold.[2] Māori iwi had battled through the courts and politically to win recognition of their right to fisheries through the Treaty of Waitangi, but it is arguable whether the original aim had been to sell quota to outsiders rather than fish for it themselves.

In the case of *Oyang 70*, various iwi, probably through an intermediary, would have received several hundred thousand dollars for leasing their Annual Catch Entitlements for the whiting, and Southern Storm Fishing would have charged for its services and taken a share of the profit. The biggest earner would have been Sajo Oyang Corporation through the catch itself. Southern blue whiting does not fetch high prices so the biggest profits could be made if wage costs were kept to a minimum.

Oyang 70 arrived at the fishing grounds on the evening of Sunday, August 15. The reproduction rituals of a fish can easily be factored on to a screen, along with Google Earth, so the Koreans quickly knew they were above southern blue whiting. They shot their first net into the heavy aggregation of fish and made a good catch, 3,000 pans. A pan (which New Zealand accident investigation authorities would later keep calling a 'pen') is the amount of fish – around 13 kilograms – that can literally be fitted on a factory pan.

This first catch of 39,000 kilograms went into storage pounds on the ship. From here it would be processed on a production line in the grandly named 'factory' on board. This would take 18 hours.

In a modern ship, the factory deck can look austere and dangerous but efficient. Conveyors quickly move the catch through the factory, where sharp knives and circular saws slice and dice the fish. These implements can also injure workers forced to balance in confined spaces no matter what the condition of the sea. If catches are going to destinations such as the European Union, tough sanitary rules must be followed. A well-organised factory will get fish snap-frozen relatively quickly.

Oyang 70's factory was neither modern nor well-organised. It was Dickensian. The rust and filth alone suggested its owners had no real interest in the final product, whatever it would be. The ship was designed for short Asian crewmen. Its low ceiling had been given as a reason why it and other Korean ships should not be forced to take observers, who were usually comparatively tall New Zealanders. The machinery was old, uncovered, corroded in places, and often failed. A modern factory deck will be wet, but on a well-run run ship this situation can be controlled. On *Oyang 70* the water flow was random, perilous and unpredictable. The chute for waste – mostly comprising heads, guts and any bycatch – was permanently and dangerously open to the ocean, just as it had been on *Insung 1*.

Fish processing consisted of H&G – head and guts. Fish went down a conveyor belt, bound for a man running an unguarded circular saw. After this man whipped off the heads, the fish moved on to a line of knife-armed men who gutted them. At the end of the line the fish were packed and frozen; later, in China and Thailand, they would be sufficiently thawed that workers could make them into some sort of marketable product.

So long as fish were coming aboard, the crew had to work: there was no prospect of rest. On the evening of August 16, *Oyang 70* shot its second trawl. This garnered a modest 700 pans, which took six hours to process.

During the trawling, the cod-end of the net – the part where the fish ended up – had been damaged, so for the trawl early the following morning a different net, bigger and longer, was set. It hauled in an impressive 3,500 pans, or 45.5 tonnes of fish, to be processed over the whole day.

Even as the men in the factory struggled to deal with the fish they had, Shin decided to shoot another net, his fourth. In the kind of operation run by Oyang, the pressure is on captains to catch as much fish as quickly as possible. Southern blue whiting assist the effort by creating large aggregations just after dark.

Just before the net was shot, the crew realised the batteries of the sensors at the net's opening were flat. When a net is in the water, the sensors measure the fish going in, showing them as coloured data on a computer screen on the bridge. This has several benefits. It is good seamanship to know the weight of the trawl so the ship remains stable. And it is essential to know when to haul in the net. An over-full net damages the catch, crushing the earlier caught fish. When this happens, much of the catch goes to waste.

There were new batteries on the bridge, and the sensors were removed so the batteries could be installed. However, Shin's haste in getting the net into the water meant the crew never got around to putting the sensors back in action. Many an old fishing-boat skipper knew by feel what was in his nets and could survive without the sensors. As would become obvious, Shin, who was driving his boat as hard as he could, was not a skilled mariner. With the boat sailing in a straight line he had no sense of its stability, and he lacked the ability to calculate the effect of having a big net trawling behind.

Shin went to bed. The crew, however, were not able to sleep: they still had tonnes of fish to process.

As dawn broke, the factory manager, Tae Won Jo, told the 30-year-old first mate, Min Su Park, that there were only two and a half tonnes of fish left to process from the earlier catches. That meant they could bring in the net. Almost as soon as they started hauling it in they knew the catch was huge. Experts in the various investigations that would follow were able to only guess at its size, placing it in the range of 70 to 130 tonnes – at least twice as large as any other catch the ship had made.

A helmsman who woke at three a.m. felt the ship had a hard heel to port. He put on his lifejacket and headed for the lifeboats. The ship was in darkness; the lights appeared to have failed. He could see a huge pile of fish on the stern deck. Another crew member who came up on deck could see fish in the net being winched on board. A huge amount of fish, the largest he'd ever seen, was falling towards the port side.

Several Indonesian crewmen on the trawl deck could see it was going to be dangerous to haul in the net any further and called to the bridge to drop it immediately. Boatswain Byeonghak Lee informed the first mate there were too many fish. An engineer on the winch warned that water was flooding into the ship. The chief engineer, Dong-Hyeon Jeong, appealed to the first mate to cut the net.

Park, not wanting to be the one to make the decision, woke the captain. Shin demanded the net keep coming aboard.

Bringing in a net is an industrial ballet requiring skill, a sequence, and the handling of a complex array of lines and winches. The net should come up the centre of the ship and be secured by lines as it is winched over the stern. The size of even the haul on *Oyang 70* that night might not have been an issue on a well-managed ship. The evidence of the crew suggests the net was too big to come in through the trawl doors, but a good skipper may have been able to cope with that.

As it came up the stern, the net slid to the port side of the vessel, putting the ship on a 15-degree list. Around 25 metres of the cod-end was now out of the water and another five metres in the water. Because the battery problem had left him without net sensors, Shin would have had little idea how big the catch was. He may not have understood that because of the earlier net damage there was a bigger net coming out of the water.

Listing increased. Ships can handle much bigger lists than that faced by *Oyang 70* if they are watertight. The Korean vessel was not watertight. Scuppers on the deck that were designed to let water flow off the ship may not have been working; given the general lack of hygiene on the ship, it is possible they were jammed with dead fish and other debris. The open scuppers were letting in the ocean.

This was bad enough but the situation was worse on the factory deck, where an offal chute dumped guts and heads straight into the ocean.

Crew would later testify that this chute was always open, no matter the weather. Water was now pouring in via the chute and flooding the factory deck. The water cascaded down another deck to the engine room. The bulkheads separating the factory from the rest of the ship, including the engine room, were not watertight, and the watertight doors that should have kept water out were wired open.

Investigators would later find evidence that suggested *Oyang 70*'s engine room may have flooded often before this. As water poured in, the first engineer's reaction was to place flattened cardboard boxes over the top of the generator to protect it, before going to see where the water was coming from. Investigators were struck that flattened boxes were so readily available in the engine room; it indicated they had been used before.

Compounding *Oyang 70*'s problems was its fuel load. Shin had expected a short voyage, believing he could quickly get as much fish as he needed. Just one tank was full and the rest only partially full. The partially full tanks were bringing what's called 'free surface effect' into play.

When a tank is full, the liquid inside it acts like a solid mass in the event of a ship listing. The tank's centre of gravity stays the same and so does not alter the ship's centre of gravity or its metacentric height – that is, the distance between its metacentre and its centre of gravity. This is important for stability.

When a tank is partly filled, the liquid inside it seeks to remain always in parallel with the waterline. Hence, the centre of gravity in the tank constantly changes. More importantly, it impacts on the ship's centre of gravity and its metacentric height. A competent skipper would have 'pressed up' his fuel tanks, using what fuel he had to fill each tank, thus limiting free surface.

To make things worse, Shin was facing the strange situation where the large, oily, squashed body of ruined fish on the deck was also creating a free surface effect. The already barely stable ship was confronting a rapidly worsening sequence of disasters.

If Shin began to realise the likely consequences of his decision to keep hauling in the net, he was not about to change his mind. Crew who went to the bridge and pleaded with him to cut the net loose would later testify that he was very concerned with saving the fish. Edwin Gonzales, *Oyang 70*'s only Filipino crew member, remembered: 'The captain want fish, not worry about the crew, we do not like captain.' According to another crew member, Imam Subekhi, the captain just stood, drinking what Subekhi thought was water. 'Everyone was scared and panicked. They tried to release the lifeboats,' he would recall.[3]

Some of the crew tried cutting the wires to the net but it was impossible. The boatswain was finally allowed to climb out on to the net astern to try and cut it open and free the last of the catch. This did little good. Nor did Shin's move to order up more power from the flooding engine room in a bid to steer the ship out of its list.

At 4.30 a.m., with water streaming into the engine room, electricity gone and the list increasing, Shin finally realised his ship was capsizing in a calm but cold and fog-covered sea. He reached for the VHF radio and made a distress call. This illustrated just how inept he was. VHF Channel 16 is short-range: he was running a real risk that no one would hear him. There are other well-understood procedures for signalling distress and throwing into action a sophisticated response.

Luckily for Shin, a New Zealand-flagged fishing boat heard the call. Skippered by Greg Lyall and carrying an all-New Zealand crew, *Amaltal Atlantis* alerted the Maritime Operations Centre in Wellington, which quickly had a Royal New Zealand Air Force P3 Orion scrambled.

As the ship developed its fatal list, an Indonesian crewman woke up and saw what was happening. He later told police he couldn't believe no deckhands had life jackets on, while all the Koreans were wearing them. 'Chief officer and other Koreans could all save themselves on one life raft, which comprised three engine people, factory supervisor, deckhand and one Chinese cook,' he said.[4]

He was also amazed there was no alarm and no instructions 'but these people [the Korean officers] ready with jackets by the bridge. Indonesians prepared the raft for them. [I] helped. No instructions. Amazed Korean deck boss not on deck or talked to deck staff. ... Indonesians took care of each other.'

Shin had said on radio that he was sinking but he did not share this news with his crew. He was apparently bent on suicide in the glorious tradition of a captain going down with his ship – and, through inaction, taking the crew with him. Two days after the disaster, Christchurch police interrogated the surviving officers and crew of the *Oyang 70*. Detective Sergeant Michael Ford, in his summary of evidence, noted that Shin had refused a lifejacket offered to him as the ship went down. 'The captain was hugging a post and crying, after drinking clear liquid from a bottle.'[5]

Oyang 70, forever attached to its enormous haul of southern blue whiting, capsized and sank with around 700 tonnes of light fuel oil aboard, in water near the Bounty Islands marine protected area. The oil would leak out slowly over the years. Captain Shin disappeared forever.

6 Cover-up

As the *Oyang 70* was going under, Maritime New Zealand's Rescue Coordination Centre alerted me to the drama. I had an advantage in covering the story as I already knew something about scandals involving foreign charter vessels, and had recently covered both the Kiribati search for *Tai Ching 21* and the sinking of *Insung 1* in the Southern Ocean.

The *Amaltal Atlantis* had quickly arrived at the scene. The sea was covered in oil and flotsam. Captain Greg Lyall ordered his ship's rigid inflatable boats lowered to round up the life rafts. Remarkably, by 6.50 a.m. they had rescued 45 survivors and soon afterwards they picked up three bodies. Another ship *Professor Aleksandrov*, a Ukrainian vessel chartered by another Nelson-based fishing company, Sealord, arrived. Although its crew spotted a body in the water, the captain refused to launch a boat, saying there was too much flotsam; the crew of *Amaltal Atlantis* picked up this body too.

Lyall, one of the best qualified skippers in New Zealand, ordered his officers to interview the *Oyang 70* survivors as soon as they had been cared for.

'We were not able to communicate with the Korean officers as they spoke no English at all,' he would tell the Coroners Court two years later.

'I noticed a very distinct separation between the Korean officers and the rest of the crew. [The Koreans] did not mix with or communicate at all with us or the crew whilst on board the vessel.'[1]

The people on board *Amaltal Atlantis* described to the court what had happened as *Oyang 70* went down. One, Andrea Haliburton, had been told by the crew that they had been hauling when the boat tipped on its side, 'no power, no lights and no alarms'. They had decided for themselves it was time to run for the life rafts, and most had had to swim to them.

Indonesian crew members informed *Amaltal Atlantis*'s Casey Seerup they had told the skipper something was wrong but 'skipper didn't care'. Other crew members were recorded as saying: 'Skipper was stupid, last catch was too much; living conditions were not good'; 'very stupid, stupid skipper'; and 'skipper was bad.'

One of the survivors asked a crew member of *Amaltal Atlantis* if they had found his brother. It was his birthday. They had found him. He was dead.

Another rescuer reported being told of the circuitous route by which the crew had reached New Zealand. After passing through Indonesian customs, they had been made to sign a contract. They had then been flown to Angola, Singapore and Australia. They had not been told what country they would be fishing out of. When the ship sank they had been fishing for only a few days. 'Skipper being greedy, too bigger bag. Crew was telling him not to haul,' the crew member said.[2] He reported that the Indonesians had not got on with the Koreans, and had ended the interview with a touching request: how could he get a job on *Amaltal Atlantis*?

The rescued men complained about the treatment they had received aboard *Oyang 70*. One told an *Amaltal Atlantis* crew member: 'Very bad agency, didn't get paid until contract finished. … Koreans always angry at crew. Not always allowed to go toilet during work hours.'

Another survivor 'got treated badly … worked as fast as they could and still got yelled at to work faster. Had only one set of work clothes to last for the whole trip.' Another spoke of being fed reject fish for all meals.

Meanwhile the crew had been working 14 hours, sometimes 18,

with only four hours' sleep. Indonesian crew members who had less than two years' service earned US$200 a trip, and those with over two years' experience $300. A crew member on *Amaltal Atlantis* reported being told: 'Food only what they catch – whiting, hoki. … No bread, no noodles or rice, just fish. Crew did not like Korean officers because they treated crew like slaves.'

As well as Captain Shin, the dead had names: Samsuri, born 1972; Taefur, born 1975; Heru Yuniarto, born 1985. Sailors Tarmidi, born 1981, and Harais, 1974, were never found.

Amaltal Atlantis arrived in Lyttelton on August 20 to an unusual police response that reflected the political power of the fishing industry and showed little concern for the survivors. Police sealed off the wharf area to prevent the media talking to the men, who were transferred to a bus and transported into Christchurch. There is only one direct route from Lyttelton to Christchurch – through a road tunnel. In an unprecedented move, police blocked off the tunnel as soon as the bus entered it, preventing the media from following.

Sajo Oyang's shell company Southern Storm Fishing was effectively using the New Zealand police as private security, something that upset individual police officers. One later telephoned and told me what had happened. Southern Storm had run the whole operation, she said, trying at all times to keep the Indonesian crewmen away from the media and get them out of the country as quickly as possible.

The Koreans and their agents kept insisting the Indonesians could make statements only for the coroner, that nothing else was legitimate. This was untrue: even when statements go to authorities, there is nothing prejudicial in telling the public what has happened.

With the police running cover, the Koreans got the crew to a motel in Christchurch and told them they could not leave it until put on a plane. Security guards stood outside.

'There was no kindness for them,' the police officer told me. 'They were treated like animals. … Those poor buggers in that motel room, they were starving. They were given one change of clothes, no money. Their food was a lot of boxes of noodles and water. They were not given anything else.'

A member of Christchurch's Indonesian community, alerted to the men's plight, managed to get help to them. 'It was obvious when we gave them food that they had not seen rice for a very long time. It was hard for them not to cry when a group of people who never knew them brought them food,' he said. 'Survivor guilt was one thing, you have to deal with that. But the fact they went through the ship sinking, and help was not available until the Talley's boat came... I've never been in a situation like this. ... It was like when you experience life and death ... they were distraught and sad.'[3]

The Korean Maritime Safety Tribunal, which had poorly investigated *Insung 1*'s sinking, conducted a similarly trivial inquiry into the sinking of *Oyang 70*: it did not even get the catch right, saying the ill-fated boat had been hauling in pollock. It determined that the ship had sunk due to maintenance issues with the scuppers, which were not working properly so water on the deck could not drain off. No one in serious maritime circles gave the tribunal's findings any credence.

Nearly two years after the disaster, I flew to Wellington to attend the inquest of Coroner Richard McElrea into the deaths of the three men whose bodies had been recovered. Fairfax Media had plenty of people in the capital who could have covered it but the story had become personal and I wanted to see if the truth would emerge.

A Korean man made an appearance to give evidence. This was extraordinary: people behind the Korean fishing industry seldom showed any kind of public face. A Nelson admiralty lawyer, Peter Dawson, had told me how frustrating it was to be in a legal battle with Korean fishing companies because they never fronted personally but were always represented by lawyers or agents.

A coroners court, however, has great powers to call evidence and witnesses, and it is an unwise person who refuses. Oh Gwan Yeol, a large young man, sweated nervously throughout his brief appearance. While he was described as the operations manager, neither his role in Southern Storm, nor whether he was directly answerable to Sajo Oyang Corporation, was clear.

Oh had heavy-hitting legal help in the form of Queen's Consul Bruce Squire. When lawyer Craig Tuck, representing the families of the dead crewmen, tried asking the Korean questions about wages and the kind of catch bonuses the captain would receive, Squire objected. The questions were 'obviously being pursued for collateral reasons'.[4] The families, he said, were using the inquest for their own motives.

My reporting was dragged in. Graeme Christie, a lawyer representing the marine classification society Lloyd's Register, told the court the question of the wages the men had received when alive was of no relevance: I had reported in *The Sunday Star-Times* that the widows had received cash benefits for the deaths from the New Zealand government's Accident Compensation Corporation.

The coroner agreed, so the only judicial inquiry into the deaths of the men from *Oyang 70* did not consider the matter of their appalling working conditions.

Coroners courts, charged with determining the causes of deaths, are among the oldest courts in the Commonwealth. By custom and long practice, the closest kin of the dead person is able to make a statement. Usually the statement is delivered in person and often orally. The three Indonesian widows had prepared a common statement. It totalled just 158 words. Bruce Squire demanded it not be tabled, saying it was being presented only for publicity purposes. Craig Tuck replied that he was presenting the views of victims.

Coroner McElrea censored the statement, ordering that words critical of the New Zealand government could not be published in the context of any report on his inquest. They could, he said, be published in another context. A specialist in media law told me it was one of the oddest suppression orders ever made, and advised me to publish the statement.

As far as I know this is the first time it has appeared in full in public.

'Thank you for the chance to speak to the Coroner's inquest,' the widows wrote.

'It was a long 20 months since the heart-wrenching loss of our loved ones yet still we do not know what happened to cause their demise.'[5]

The next sentence, which I was forbidden from reporting at the time, read: 'The NZ government has not told us anything. We trusted NZ government to keep our men safe because NZ is a safe country.'

The statement went on: 'We know fishing is hard work but we believed the fishing company would look after our men.'

The next three sentences were suppressed: 'Now we are without their income in our family. We must struggle every day and hold our dignity. Since we learned they should be paid NZ minimum wage it is worse because we never received the minimum wage even to this day.'

The next part was approved for publication: 'We are very grateful to our representatives and Accident Compensation Corporation who finally achieved some money for us after this time. We ask that New Zealand government and the fishing company do not forget us.'

The coroner's verdict was nearly a year in coming.

7 Slavery

Life moved on as judicial and political wheels ground. I had been working the story about the conditions on foreign charter vessels fishing for New Zealand quota for a long time, but the core of it had yet to go to the New Zealand public. Many people, particularly politicians, were well aware of it. One leading Labour Party member of parliament – safely in opposition – told me he had heard all the horror stories but been urged to keep silent. As I discovered later, some politicians had accepted campaign funding from the fishing industry. It behoved them to look the other way.

Readers of *The Sunday Star-Times* on April 3, 2011 did not have that option. 'Slavery at Sea' dominated the front page, along with the blurb: 'Alerted to terrible conditions on foreign fishing vessels after nearly 30 people lost their lives, Michael Field began asking questions.'[1]

The newspaper had given the exposé maximum coverage: it ran on the front and inside pages. Readers were informed that the ships were 'little more than floating sweatshops', with crews coming from low-wage countries and unscrupulous agents taking half the crews' already abysmal pay.

We listed conditions allegedly found. Boats had kept working after men from them had drowned. Crews had been slapped, punched, and hit with fish and hammers. Lifeboats had not been in working condition. Crew health problems had been ignored, and men had had to work for days on end without breaks. It was standard for them to have their passports confiscated so they couldn't leave.

We revealed that the New Zealand government had known of the problems for some time. Papers obtained under the Official Information Act showed it had been aware of how little the men were being paid and the role of agents. It was clear it was allowing fishermen from poor countries to be badly treated while the companies who hired them exploited large parts of the country's fishery. The foreign charter catch, worth $300 million a year, was meanwhile being marketed as 'Produce of New Zealand'.

In the papers we obtained, a government official admitted that crewmen had told him they had never worked in such terrible conditions. 'If these tales are correct, and I have no reason to doubt them, the conditions amount to little more than "sweatshop" ones,' he had stated. 'If anyone stands against the abuse, it has been known for them to be taken to a cabin and beaten. Individuals are reluctant to draw attention to themselves.'

On one boat a crewman with tuberculosis had received no help until fishing ended. When finally put ashore, he was hospitalised for three weeks. A fellow crewman with appendicitis had received no medical help at all.

On another ship a man working in a freezer had suffered frostbite that was bad enough to require hospital treatment. When he returned to the vessel he had been made to remove his dressings and get on with work.

Talley's Group chief executive Peter Talley went on record as saying the government had dealt with the issue only to the extent of setting basic standards for New Zealand observers put aboard vessels. 'They do not care about the Filipinos, Indonesians and Ukrainians,' he told *The Sunday Star-Times*.[2]

Long before the coroner had held an inquest into the sinking of *Oyang 70,* Talley had told me that when the *Amaltal Atlantis* rescued the survivors it discovered that nearby foreign charter vessels had ignored the ship's distress calls. 'They wouldn't knock off fishing,' he said. 'We were the only ones who put man-overboard rafts into the sea.'

Talley described New Zealand fishing as being like the Wild West and getting worse. 'Because of the higher price of fuel around the world, more of these boats are getting displaced and New Zealand is ending up as the junkyard for the fleets. I think I can say without fear of contradiction that nothing has improved one bit, and in many ways things have got markedly worse.'

Talley's company has eleven ships, pays its crews New Zealand rates, and still makes a profit. 'You would be amazed at the money the boys on our boats are making,' Talley told me. Deckhands and factory workers earn between $40,000 and $80,000 a year.

Unscrupulous operators were, he said, using a model that didn't require access to capital, catch entitlements, or ownership of processing factories or fishing boats. 'People are knowingly saying, "I don't pay the crew, I just charter the boat. The crew is provided by an agent and the agent tells me he is paying full wages."'

This was wrecking the viability of New Zealand fishing. 'All you need is an office, a secretary, a tough lawyer, and you are in business. But it is not good for New Zealand.'

The *Sunday Star-Times* story included the information that an Auckland law student, Jennifer Devlin, had written in *Australian and New Zealand Maritime Law Journal* about an incident in which ten Indonesian fishermen had escaped from a Korean vessel, *Sky 75*, in Nelson. 'They were fed rotten meat and vegetables, told to shower by standing on deck in the waves, made to continue working when sick or injured, and constantly beaten. They endured all this for wages of US$200 a month – wages that weren't being paid,' Devlin wrote.

On their return to Lyttelton, the *Oyang 70* survivors had been quickly moved out of New Zealand on a pre-dawn flight to Singapore. Most did not get the wages due to them. Sajo Oyang and Southern Storm must have presumed that, like all the other fishing scandals, this too would quickly be forgotten. Less than 12 hours after *Oyang 70* sank, Southern Storm went online to the Companies Office and changed its shareholding structure. On the same day that six men were killed and survivors were being hosed down and given their first decent meal, the corporate response was to take the time to change the ownership of a small shell company.

A month earlier Sajo Oyang had taken 450 of the 1,000 shares in Southern Storm Fishing owned by Hyun Gwan Choi of 84 Stanley Road, Christchurch. Now the registration was changed to note that Sajo Oyang had moved the shares back to Choi. A casual look at the Southern Storm Fishing's company registration would show no Oyang link, although anybody familiar with the system would know to look at all the earlier documents.

It was a strange move. In the unlikely event of a prosecution, the change in share structure would make no difference. When I made this point in a couple of stories, Pete Dawson, the agent looking after Oyang vessels on behalf of Southern Storm and its sidekick Fisheries Consultancy (NZ) Ltd, launched into me, saying I did not understand company law and structures.

He was wrong, as he was about many things. What I was seeing in the shuffling at the Companies Office was the same kind of behaviour I had observed when investigating the way in which shell companies were created in New Zealand for all kind of nefarious activities. I had covered an Auckland shell company chartering a Georgian plane to fly North Korean arms to Iran, and shown how Russian mafia and Mexican drug cartels used New Zealand shell companies to launder money. Sajo Oyang's modus operandi looked like that of so many other dodgy companies. The world of shell companies and the associated criminality was something with which I had become very familiar: shell companies and reflagging have always been crucial aspects of global fishing, a way to hide the truth.

Shortly afterwards I was leaked a memo written by Kyung-so Moon, Sajo Oyang's 'chief of New Zealand base'. He had taken the step of reducing problems with a share transfer, he told his bosses in Korea. 'Hyun Choi sold the stake of Southern Storm in his family trust. The stake was shared and transferred to the wife's maiden name. Additionally, Hyun Choi withdraws its place as the director of his company; his wife's maiden name will take over the position.'[3]

As I chased down the story behind the sinking of the *Oyang 70,* it was a struggle to elicit any kind of Korean response. I tried to get comment from officials in a number of companies, including Sajo Oyang, on various allegations of labour abuse and human rights. They ignored me. I approached the Korean Embassy in Wellington. A senior diplomat, Tony Sung, sent me an email, the likes of which I had never seen before in my 40-year career as a journalist.

'We hope,' Sung wrote, 'that in any context neither the Korean Embassy nor the Korean Government will be referred to in any article about these issues and we wish such hopes will be respected by you. From now on, any further correspondence from you on these issues will not be acknowledged or responded to. Also, the content of this email must not be disclosed, quoted or referred to in any form whatsoever.'[4]

The embassy later developed an interesting strategy of giving exclusive interviews to New Zealand's TV3. The relationship with the channel developed to the point where the Korean government gave a news crew a free trip to Korea to see the glories of its economy.

While I was reporting on all this, heavy-hitters at the University of Auckland Business School were using the school's considerable resources to investigate foreign charter vessels fishing for New Zealand quota. Christina Stringer, a senior lecturer in the Department of Management and International Business, and Glenn Simmons, a doctoral student and former law enforcement officer, went to Indonesia and tracked down the survivors of *Oyang 70* and the families of the men who had died. They were followed by Daren Coulston, a human rights advocate and former deep-sea fisherman. Their subsequent report 'Not in New Zealand's waters, surely?' was a devastating critique of foreign-flag fishing.

Stringer and Simmons had met with a young widow and mother they called Eula – not her real name. Eula had helped her husband get a job on *Oyang 70* through an agent; promised earnings were US$280 a month. She had had to sell her wedding present, a gold necklace, to pay the agent's five million rupiah ($440) fee. Under the deal, the agent would pay 1.5 million rupiah ($130) each month to Eula and retain the balance until Mohammad returned home at the end of his two-year contract. The agent would take up to 50 percent as a fee.

After her husband died on *Oyang 70*, Eula went to Jakarta to get the insurance money she believed she was entitled to. One of her friends told Glenn Simmons, 'They said husband's insurance money has not come from the Korean agent, and if you want to get insurance money you must sleep with the director of the agency for a few days. After [she came] back she cried loudly and could not speak. Then she asked where should she complain, whether there is justice and righteousness on earth.'[5]

Eventually, with the help of Coulston, Stringer and Simmons, Eula and the other widows had filed a claim with New Zealand's Accident Compensation Corporation. Normally, crew of a foreign-chartered vessel would not get payment from ACC as they are not New Zealand tax residents. In the end, however, the corporation paid out. Neither the Korean company nor the Indonesian agent paid anything to the widows of the men who had died working for them.

As news of the plight of charter vessel crews got more media exposure, supporters of the companies involved mounted a series of campaigns to destroy the credibility of those working to end the practice. A man called John Albert Situmeang, claiming to be the chair of something called 'Indonesian Fisherman Federation', emailed dozens of people, claiming that Coulston was a 'robber to extract money from Indonesian fishermen' and that others involved in the campaign were taking 40 percent of any money raised for the struggle. 'The main goal you really want to rob money Korean fishing vessel owners,' he wrote.[6]

I tried talking to him but he was evasive and unable to provide any kind of evidence. Coulston, who denied the claims, said Situmeang wasn't even the man he said he was. Coulston had worked on foreign fishing vessels and skippered vessels fishing in New Zealand, the Solomon Islands, Australia and international waters. He had also owned vessels, which gave him significant insight into crew management and vessel operations. It was his view that researching the payments of crews was the way to expose the deception and fraud, something he was committed to doing.

'Forced labour is evil slavery and at no point will I give up, ditch the widows and crews, let the vessel owners and New Zealand fishing company guarantors get away with it, and allow the ineffectual government to swim free,' he told me.

The publicity I had generated with my first stories, which had been followed up by other local and international media, had drawn attention to the University of Auckland Business School, and in particular to Christina Stringer and Glenn Simmons. Some commentators saw the revelations as a fine example of media and academia cooperating. In fact it was mostly a coincidence.

My interest in fishing conditions had come about as a result of years covering the Pacific and noting, among other things, the appalling state of crews on boats moored in Suva Harbour. Christina Stringer's interest sprang from a research project she had undertaken from 2008 to 2010 for the Ministry of Fisheries, examining the extent to which fish caught in New Zealand waters were exported to China for added-value processing before being sold in key international markets. Simmons had been her research assistant on the project.

As an unintended consequence, the pair had found that some foreign charter vessels fishing on behalf of New Zealand companies and agents in New Zealand's exclusive economic zone resorted to slavery aboard their vessels. Like me, they became aware of what was going on with Korean boats in particular. And, like me, they experienced threats and intimidation. One night they had dinner with crew members of a Korean ship at a restaurant in Auckland. When they left the restaurant, the crew members' former New Zealand employers and three of their associates were waiting outside. These men sought to intimidate the group to the point where they were fearful for the crew members and themselves. 'We were stalked, photographed, and our vehicles followed,' Stringer recalls. 'We were forced to take a series of evasive manoeuvres in order to lose them.'[7]

Meanwhile, private investigators were put to work looking into my background and scrutinising Simmons and Stringer in an effort to find out how we operated.

8 Verdict

Coroner McElrea produced his 44-page findings nearly a year after the hearings in Wellington and two and a half years after the *Oyang 70* sank. Coroners' findings can be critically important in public understanding of a tragedy and they often take on heroic significance. McElrea produced a bland document, poorly written, disconnected, and reflective of the way lawyers had fought over it.

He had to report on only three of the six men killed: Heru Yuniarto, Samsuri and Taefur. Their bodies had been recovered and post-mortems showed they had all drowned. The other men, including Captain Shin, had been lost at sea. He had to hand a report by New Zealand's Transport Accident Investigation Commission and one by the Korean Maritime Safety Tribunal. The latter was quickly dismissed: 'No reliance is placed on this undated and unsigned document produced on plain paper without any official setting.'

McElrea decided early on that the size of the catch had been a vital factor in the sinking. However, Robert Leyden, the principal surveyor for the ship classification society Germanischer Lloyd NZ, had given evidence that a properly watertight ship could have easily survived a heavy listing. The logical conclusion was that *Oyang 70* was not watertight.

The coroner highlighted the key dates and events in the last part of the ship's life. In November 2008 the vessel had been dry-docked in Lyttelton and subject to a full fabric maintenance and hull survey, which it had passed. In December 2009, Maritime New Zealand had conducted an inspection and found 15 items that needed fixing.

A month later, Lloyd's Register Asia had carried out a 'survey afloat' for a new safe ship management certificate. Thomas Battrick, the inspector, had checked such items as muster lists and instructions for lifeboat launching. He had interviewed the master and another deck officer as to emergency and other procedures. He had satisfied himself about such matters as the servicing of fire appliances and life rafts, and the correct stowage of life rafts. He decided the ship fell narrowly within the category of 'low risk'. He was satisfied the vessel's owner and skipper were aware of 'all applicable maritime rules' and showed acceptance of and compliance with them. He signed a Fit for Purpose Certificate, indicating compliance with 'applicable maritime rules, marine protection rules, Resource Management (Marine Pollution) Regulations 1998, and Health and Safety in Employment Act 1992'.[1] He had, McElrea said, carried out his task in a reasonable manner.

In March and April 2010, a Ministry of Fisheries observer had been on board the ship during fishing operations. In July 2010 a Maritime New Zealand inspector had again checked the ship and let it sail.

The ship had left Port Chalmers with almost empty fish holds. McElrea noted that it had 'marginal stability'. The Transport Accident Investigation Commission had pointed to the 'free surface' effect in the partially full tanks. The coroner agreed: 'A prudent master will ensure that fuel, ballast and freshwater tanks are "pressed up" before departure from port.'[2]

Expert witness Andrew Leachman, a former ship's captain, had said that pressing up required tanks to be 100 percent full, with no free surface effect possible. Crew had said the ship felt unbalanced. 'The master, for commercial reasons, may well have chosen to sail with the vessel in a "tender" state and with marginal stability,' the coroner noted.

McElrea's report included accounts from crew members saying the 120 tonnes of fish was the biggest haul they had ever seen. According to the helmsman, there were 'three times more fish than usual in the nets'. The mouth opening of the final trawl net would have been the size of six tennis courts: 140 metres wide and 60 metres high.

Leachman was critical of the decision to fish at all that night: 'To attempt to catch more fish when the factory storage space was congested indicates poor judgement by Shin. An experienced Indonesian crewman who had worked for Oyang for a decade submitted that "this company usually loads too much fish".'

Five or six tonnes of the 120-tonne catch would have been on the 15-metre ramp at the stern of the boat, while the rest trailed in the sea like 'a big silver sausage', as Leachman put it. The weight would have caused *Oyang 70* to squat in the water, its bulbous bow out of the sea.

The *Amaltal Atlantis* skipper Greg Lyall had told the coroner of seeing other Korean ships engaged in the same poor fishing practice. He himself would 'never risk my vessel hauling a bag of that size and in those conditions': he would be immediately dismissed as skipper. Robert Leyden was also of little doubt that the decision to pull in such a large bag 'would certainly have contributed to the vessel's lack of stability'.

There was a huge amount of discussion in the coroner's report about how enough water to sink the vessel could have entered but really this barely mattered: the ship was in such a poor state that if one way in had been fixed, another two or more would still have led to the sinking. *Oyang 70* was equipped with doors on its waste chutes that were to be closed when not in use. A factory hand who worked on *Oyang 70* for the 20 months before it sank had confirmed the doors were never closed. Another worker had said they did close the chutes in heavy seas but water still got in. There was no evidence that on the morning of the sinking anybody had tried to close the chutes.

At the inquest a great deal of time had been spent studying the ship's scuppers and discharge sumps. These are basic devices that are supposed to let water on a deck flow back into the sea without letting sea water into the ship. For ships with large, enclosed decks and small freeboard – that is, decks not very far above water level – the scuppers and valves are crucial. Rubbish can block them but this is not common and they are easily cleared. 'In all my years at sea, the only time I have ever encountered issues with non-return valves is when debris is stuck in them,' a seagoing skipper, Allan Dillon, told the court.

Sajo Oyang Corporation's lawyers were keen to blame the scuppers, if only because this would cover up their fishing operations and the boat's other faults, but the coroner found the evidence that the scuppers were to blame 'very confined'. Nicholas Jessop of Lyttelton Engineering had described the valves as 'simple robust mechanisms'; it was rare to find one not working. Since a faulty or jammed 'storm valve', as Jessop called them, would let sea water flow both ways, such a fault would have been readily noticed before the sinking and easily fixed.

The ship's factory manager, Tae Won Jo, had told police 'water was coming through drains on the side of the ship'. McElrea concluded that, while some water may have entered through malfunctioning scuppers, most had come in through the offal chute.

Making matters worse, the fish holds were open for the snap-frozen fish, so the water flowed downward from the factory. The engine room – which, as for survival in any ship, should have been weathertight and watertight – was below the factory deck. As water flooded in, Harais tried to close the engine-room door. He failed, paying with his life.

The chief engineer claimed he had closed the engine-room door. No one agreed with him. Andrew Leachman told the court that the watertight doors down to the engine room should never have been left open. The safety culture aboard the *Oyang 70* was 'not impressive'.

Captain Shin was 41. He had been chief officer on *Oyang 70* for seven months and master for nine. This did not add up to much knowledge.

Coroner McElrea described the scene leading up to the sinking. A winch operator has had just three hours' rest after working 27 hours straight.

At three in the morning he is woken and told to bring up the net. As he does so he feels the net unbalance the ship. Someone from the engine room comes to the winch room and says the water in the engine room is already at knee-level. Shin is told, and asked if the net should be let go. He does not want to lose all the catch so he instructs someone to slowly cut the net and let some fish go while retaining the bulk. Before this can be done, the net tips the ship. The crew are already out on the deck getting ready to abandon ship.

Shin keeps trying to persuade crew members to centre the net on the deck – something that, given the weight of the catch, is clearly impossible. As the lifeboats are put into the water there are no further instructions from the captain.

It had been tentatively argued before the coroner that there was something cultural about Koreans and their leaders that meant they could not take crucial decisions. Cutting the net free, several expert maritime witnesses argued, would have been a gross loss of face for Shin. Around 100 tonnes of dumped southern blue whiting would have been worth around $13,000 – perhaps as much as $17,000 – in Annual Catch Entitlement. Dumping that amount of fish might also attract a criminal penalty, although Shin could have argued it had been necessary in order to save the ship.

Then there was the operating cost of the ship, around $20,000 a day, which still had to be paid regardless whether or not it caught fish, and the replacement cost of a lost net, around $200,000.

In theory, of course, Shin would not have needed to lose the net. Once he knew the size of the catch he could have reduced speed, launched the rescue boat, and directed his deckhands to cut away ropes and open up the cod-end knot while the net was slowly towed behind the vessel. A slow steam ahead would have flushed out the catch. There was one hitch: the rescue boat did not work.

Some attention was paid to Maritime New Zealand's inspection of the *Oyang 70* in July 2010. The surveyor, Robert Leyden, had had 'serious doubts' about the value of these inspections, not least because the inspection team did not seem to have determined the actual condition of the ship, nor the competency of its crew. Inspector Peter Dryden, who had gone aboard with Southern Storm Fishing's Pete Dawson, had carried out only a visual inspection. Dryden told the coroner that a Korean officer had trailed around with them, but he had dealt with him 'to a lesser degree' because he spoke little English.

On his first inspection the 15 deficiencies he had found included 'factory watertight door missing'. He explained to the coroner that the door was not missing as such, but the sealing on it was not correct. On another occasion the life rafts were not attached to the lines that would let them go if the ship went down. The way they were fixed meant they would go down with the ship if it sank.

Dryden had come up with a rating that indicated *Oyang 70* was high risk. One of the boxes he ticked showed that Sajo Oyang Corporation, the owner, and Shin, the skipper, 'showed awareness of all applicable maritime rules but showed no evidence of acceptance or compliance'.[3] This was hard to reconcile with the statement of the Korean owners that before *Oyang 70* sailed all issues had been fixed.

After Dryden had paced the decks of *Oyang 70* he had written on his checklist 'Vessel floats'. Even in a courtroom filled with big-billing lawyers the cynicism of that statement had caused unease, especially as elsewhere Dryden had described the ship as an 'old vessel'. In evidence he added that 'FCVs [foreign charter vessels] are all old vessels' and he considered *Oyang 70* to be 'at the mid-range to higher end in terms of her seaworthiness'. Despite Dryden's casual remarks about the ship's seaworthiness, McElrea noted that his action in ensuring that life rafts were correctly linked to the ship 'may have resulted in saving of life'.

McElrea had been anxious during the hearing not to explore in too great detail the abuse of the crew, something I and others felt was crucial to the ship's sinking, but in his written finding he came close to recognising it. Ministry of Fisheries' observers had, he noted, filed reports over several years, including one about conditions under Shin. The unsafe working conditions on the *Oyang 70* would not have been tolerated on any New Zealand ship, and were 'relevant to the sinking of the vessel. The safety culture and standards on the *Oyang 70* are reflected in the manner in which those in key positions … treated other crew members.'[4]

Fundamental to caring for crew on a ship is a safety plan to get off in an emergency. There was no such plan on *Oyang 70*, and no emergency alarm. The ship's rescue boat, which could have held 50 people, had no outboard motor attached. And no one could explain why none of the 68 immersion suits on board had been used. Given the sea's temperature – seven degrees Celsius – these suits would have allowed the men to survive in the water for six hours, as opposed to only two hours in normal clothing.

Instead it had been a matter of every man for himself. 'Factory personnel were bizarrely left processing fish until they were in water of a metre's depth, and left their work stations at their own initiative shortly before the ship's electricity failed and the vessel rolled over,' the coroner noted. 'Admirably, there is evidence of Indonesian-Filipino crew members looking after each other. There is no evidence of qualified personnel carrying out defined tasks in ensuring the orderly evacuation of crew.'

There was a brief window for this orderly evacuation to occur – survivors' estimates varied between ten and 30 minutes. During this time, Shin was using the loudspeaker to give instructions in Korean as to how to handle the massive catch. It was only towards the end that he issued instructions to go to life rafts – and they were also in Korean.

There was damning criticism in the coroner's report about Shin having sent a distress call on shortwave VHF radio, when he should have issued it on his two 406 MHz emergency position-indicating radio beacons. Distress calls should also have gone out on the Digital Selective Calling service on high frequency, medium frequency or Inmarsat satellite phone. DSC messages are pre-recorded distress signals that give a ship's name and number and its coordinates. The message is sent repeatedly until acknowledgement is received from a coastal station. Other ships can detect the DSC, and if no acknowledgement is heard they resend the message to a coastal station.

Through a piece of luck, *Amaltal Atlantis* had happened to be in the neighbourhood and arrived 53 minutes after Shin issued the distress call. By then *Oyang 70* had disappeared beneath the waves. Without *Amaltal Atlantis*, most of the men on *Oyang 70* would have died, and for hours no one would have noticed.

Coroners' reports end with findings and, on occasions, recommendations. The recommendations are not binding on anybody; the fact they are seldom acted upon is a source of irritation to coroners.

McElrea found that the primary reason for the *Oyang 70* foundering was mismanagement by its master, Hyoniki Shin. 'His attempt to haul a 120-tonne bag of fish on to the trawl deck of a vessel with marginal stability set in place a catastrophic and sudden chain of events that he and his Korean command personnel were unable to counter.'

He found that Shin had failed to react in a professional manner, and that personnel on board, other than those of Korean nationality, had little or no opportunity to influence the outcome, other than to assist each other.

'The vessel was not run in an orderly fashion and there was a poor safety culture,' the coroner concluded. His finding: 'Heru Yuniarto (born 2 June 1985), late of Indonesia, Sailor, died of drowning on 18 August 2010 when the Korean-registered stern trawler *Oyang 70* sank in the South Pacific Ocean approximately 400 nautical miles east of the South Island of New Zealand, near the Bounty Islands.'[5]

There were similar findings for Samsuri (born 20 April 1972) and Taefur (born 21 August 1975).

New Zealand regulatory authorities emerged unblemished. They had, the coroner said, inspected the ship according to existing rules and passed it.

■ ANTARCTIC TOOTHFISHING

Danger and death in the south's cruel seas

A South Korean fishing boat capsized in the Ross Sea in December 2010, killing half its crew. DEIDRE MUSSEN has obtained a South Korean report into the tragedy.

Rescue: The Insung No 7 arrives at Bluff in December 2010 with survivors and five bodies from the sinking of its sister ship Insung No 1. The crew were ill-prepared for emergency. Photo: FAIRFAX NZ

The remote Southern Ocean is a cruel place for a ship to capsize. Freezing waters will kill people quickly, if the shock does not stop your heart immediately.

Only half the 42-strong multinational crew on the South Korean longline toothfishing boat Insung No 1 survived after it suddenly sank on December 13, 2010, about 1900 nautical miles (1800 kilometres) north of McMurdo Station in the Ross Sea.

A day later, Rescue Co-ordination Centre New Zealand said survival times for missing crew members in such icy water would have been short, and it called off the search.

"The medical advice is that those who did not suffer cardiac arrest on entering the water would likely be unconscious after one hour, and unable to be resuscitated after two hours," mission co-ordinator Dave Wilson said at the time.

"Unfortunately, the Southern Ocean is an extremely unforgiving environment."

He said the boat sank quickly and the crew had to abandon ship without time to don emergency gear.

This week, The Press obtained a report into the tragedy, which showed the odds were stacked against the men's survival if disaster struck, and raised questions about toothfishing controls in the Ross Sea.

South Korean delegates lodged the report at last October's annual meeting of the Commission for the Conservation of Antarctic Marine Living Resources (CCAMLR),

which controls fishing in the Southern Ocean.

It blamed several safety failings, revealing that the Insung No 1 capsized after three-metre waves hit its starboard side, flooding the upper deck through its open net-hauler shutter, where fishing nets were pulled in.

"As the water flew in, the vessel began to be tilted to the right and by the time the gear stowage and engine room were submerged, the vessel lost its stability and capsized," the report said.

Language barriers between the multinational crew members also harmed their survival chances.

The boat had 40 crew and two observers from six countries – South Korea, China, Indonesia, Vietnam, the Philippines and Russia. This made swift and efficient communication between them difficult, the report said.

Many were hired through six agencies and the boat's owner, Insung Corporation, had only partial knowledge of their names and nationalities.

The ship's safety guidebook was written only in Korean; emergency and lifeboat instructions were only in Korean and English.

"This left the crew unprepared for accidents like this one," the report said.

The boat-master's response was insufficient once crew warned him at 6am that the 58-metre boat was rolling and pitching in rough seas, with water coming in through the net-hauler opening.

He failed to close the opening to stop more water flowing in, and did not check that the electric water pump was properly functioning. At the time of the

accident, the pump did not work.

Instead, the master ordered the crew to move fuel from the starboard to port side to rebalance the boat, but it failed to help.

Within 30 minutes, the ship was significantly tilting starboard (to the right). The master turned its direction to the right so the bow faced the wind and waves.

Unfortunately, it made the Insung tilt even faster to the right and it capsized from the stern

about 6.25am, less than half an hour from the first warning, the report said.

"It is considered that these harsh weather conditions made it difficult for the master and crew members to effectively deal with the situation," it said.

"Also, the water temperature at the time of the accident was between 0 and -1 degree Celsius, freezing enough to cause hypothermia for the victims."

The South Korean longliner Hongjin 707 was nearby and immediately went to the stricken boat's rescue, plucking 11 crew members from two lifeboats, and nine men, plus one Russian observer, from the water.

"Among the 42 people, five crew members lost their lives and 16 (including the vessel master and a South Korean observer) are still missing," the report said.

Its numbers were at odds with

Saved: Three sailors escape death thanks to tree trunks. They are an unnamed Indonesian sailor, left, and Vietnamese sailors Nguyen Mau Hien and Tran Dinh Khanh, who were photographed by fellow Vietnamese Le Quang Ruc. Photo: VietNamNet

The water temperature at the time of the accident was between 0 and -1 degree Celsius, freezing enough to cause hypothermia for the victims.

Report into the sinking of the Insung No 1
Only half the 42 crew survived the 2010 disaster

figures from Rescue Co-ordination Centre New Zealand, which said 20 people died after the boat sank, with five bodies retrieved and 17 crew missing.

Among Korea Maritime Safety tribunal recommendations were that all boats in the Southern Ocean should be fully prepared for harsh weather "since high waves and heavy winds in the area often cause vessels to roll and pitch", the report said.

All shutters on the sides of boats should be tightly closed when not operating and safety-related materials should be written in languages so all crew members could understand it.

Though it was unclear how the rescue unfolded, the recommendations said victims in the water should take priority over those on boats "since cold water can easily cause hypothermia".

New Zealand's Chief Coroner, Judge Neil MacLean, was still investigating the men's deaths, a spokesman for the chief coroner's office said. It had yet to be decided whether a coroner's hearing would be held.

The spokesman said an inquest into the deaths of six crewmen on the South Korean-flagged fishing boat Oyang 70 was expected to go

ahead by the middle of this year.

In August 18, 2010, the ship, chartered by New Zealand company Southern Storm Fishing, sank about 730km east of Dunedin. Rescue boats saved 45 crew, but three Indonesian crew drowned and its South Korean captain, plus two other Indonesians, were missing, presumed dead.

On Wednesday, the South Korean toothfish longliner Jeong Woo 1's accommodation block ignited in the Ross Sea, killing three Vietnamese crew and injuring seven others, including three with serious burns.

The 37 survivors were rescued by toothfishing boats near the site 600km north-northeast of McMurdo Station on Wednesday.

Four weeks ago, the Russian fishing boat Sparta was stranded in the Ross Sea after it struck submerged ice and holed its hull.

It was stuck there for 12 days while repairs were made and it arrived at Port Nelson last Tuesday.

For the year starting last December 1, 29 boats from nine countries were licensed to catch Patagonian or Antarctic toothfish in waters covered by the convention, including seven New Zealand boats.

A report by *The Press* journalist Deidre Mussen on the sinking of toothfishing boat *Insung No. 1* in December 2010. Only half the 42-strong crew survived the disaster after jumping into the freezing water. *Fairfax Media New Zealand*

Conditions on board a Korean foreign charter vessel fishing for New Zealand quota. These photos were obtained from the Ministry for Primary Industries under the Official Information Act. The ministry refused to say which boat they were from.

LEFT TOP Fish storage pound.

LEFT Fish processing factory.

ABOVE Conveyor belt for heading and gutting.

Photographs: Ministry for Primary Industries

Crew conditions on the unnamed Korean fishing boat.

LEFT TOP Kitchen area.

FAR LEFT Crew toilet.

LEFT Crew bath. In many foreign charter vessels only cold seawater is available for washing.

ABOVE Crew quarters. On some fishing boats, crew members are trapped at sea in terrible conditions for months, sometimes years, visiting ports only rarely.

Photographs: Ministry for Primary Industries

Fish being sorted for dumping on board a Korean boat fishing for New Zealand quota. On its first voyage from New Zealand, Southern Storm Fishing's trawler *Oyang 75* dumped fish worth around NZ$755,000; on its second voyage it dumped fish worth around $1.4 million. Dumping is illegal. *Oyang 75*'s captain and officers were fined a total of over $420,000 for this and other offences. In this photo obtained from the Ministry for Primary Industries, the crewman's face has been masked to conceal his identity.

Ministry for Primary Industries

On a Korean fishing boat visiting Bluff in the South Island of New Zealand, a crew member welds with only a plastic bag on his head as protection from sparks, February 2010. *Harry Holland/Maritime Union of New Zealand*

Foreign charter vessel *Oyang 75* in New Zealand's Lyttelton Harbour, June 2011. After this ship had done two trips fishing for hoki for Southern Storm Fishing, all 32 Indonesian crewmen walked off, alleging abuse by the Korean officers. In May 2014, assault charges against the officers were pending with the Korean Coast Guard Police. *Fairfax Media New Zealand*

9 Tangaroa's bounty

I n 2011 New Zealand was turned into a 'stadium of four million' to host the Rugby World Cup. Aside from the rugby fields themselves, one of the gathering places was Auckland's newly revived waterfront, replete with restaurants, venues, walkways and exhibits. Right outside the shiny new Viaduct Events Centre, unnoticed by most, was a ship of shame: 34-year-old Korean-flagged *Shin Ji*.

Like many vessels of its ilk, *Shin Ji* had borne many names. Built as a longliner to go after tuna, it had been known as *Zensei Maru 23*, *Dong Won 627* and *King Fook 88*. Now it was rusted and decaying, a death ship. The story of *Shin Ji* went to the heart of the relationship between neo-slave vessels in New Zealand and indigenous Māori.

New Zealand's founding story is based on fishing. Polynesian demigod Māui went fishing with his brothers and hauled up the island that is today New Zealand's most populated: Te Ika-a-Māui, the fish of Māui, more commonly known by its prosaic name, North Island.

In 1840 most Māori tribes living in New Zealand signed the Treaty of Waitangi with the English monarchy, ceding 'to Her Majesty the Queen of England absolutely and without reservation' sovereignty over the country. In the English-language version, the queen also 'confirms and guarantees to the Chiefs and Tribes of New Zealand and to the respective families and individuals thereof the full exclusive and undisturbed possession of their Lands and Estates Forest Fisheries'.

Fisheries were important. The country had no land-based mammals other than a long-tailed bat, and so for Māori the rivers and seas provided the food for survival. Of necessity these Polynesian people had a sophisticated understanding of the coast. They also had some knowledge of the deeper ocean, although the evidence suggests they were not reliant on it for food.

The first recorded instance of trading in fish between European and Māori was by Captain James Cook on October 15, 1769. 'At 8 am … some fishing boats came off to us and sold us some stinking fish,' Cook wrote in his log. 'However it was such as they had and we were glad to enter traffick with them upon any terms. … in a very short time they returned again and one of the fishing boats came along side and offer'd us some more fish.'[1] The 'stinking fish' was probably dried shark, a food popular with Māori in the area.

Joseph Banks, the botanist on Cook's *Endeavour*, thought fishing to be 'the chief business' of Māori they visited in the Bay of Islands. Māori had 'a little laught' at the *Endeavour*'s seine net and produced one of theirs. '[It] was 5 fathom [nine metres] deep and its length we could only guess, as it was not stretched out, but it could not from its bulk be less than 4 or 500 fathoms [730–900 metres].'[2]

The first European settlers were amazed to see the high standard of fishing and its ingenuity and industry. James Busby, British Resident at the Bay of Islands and co-author of the Treaty of Waitangi, was well aware of the importance of fishing to Māori: it was he who insisted the words 'their fisheries' be put in the treaty.

The missionary William Colenso noted that Māori 'were very great consumers of fish'. They knew the proper seasons, the best places and the best manner in which to take them; 'sometimes they would go in large canoes to the deep-sea fishing, five to ten miles from the shore, or with large drag nets [would ensnare] great numbers of … those fish which swim in shoals.'[3]

In 1843 the *New Zealand Journal*, published in London, stated that a fishing trade in New Zealand held 'bright prospects … and there would be the assistance of the natives in furnishing the nets and manning the boats…' In 1886 an observer had never seen 'the sea more alive with fish than I have seen around the Hen and Chickens, where in fine weather were constant shoals of fish, stretching as far as the eye could see in every direction'.[4]

With the expanding European population and the ensuing wars, Māori, including those of the north, lost control of their land and fisheries. There were those, such as Ngāti Whātua leader Āpihai Te Kawau, who objected. 'It was only the land I gave over to the Pākehā, the sea I never gave, and therefore, the sea belongs to me.'[5]

Such objections were in vain. A century later in 1977, when the New Zealand government passed a law declaring a 200 nautical mile (about 230 mile) exclusive economic zone – the fourth in the world (or perhaps the fifth: the facts are something of a moveable feast) – it was widely assumed this had no particular bearing on Māori rights. This assumption was to lead to lengthy hearings before the Waitangi Tribunal, a body set up in 1975 to help resolve disputed treaty claims.

The tribunal would make two hugely significant findings. In 1988 it reported on the fishing interests of Muriwhenua, the original people of New Zealand's far north. 'The libraries of their minds are replete with an enormous treasure trove of ancient practices, customs, beliefs and laws telling of the huge reliance upon the seas in days gone by,' the tribunal found.[6] Several hundred fishing grounds, up to 40 kilometres (25 miles) out to sea, were named and identified in detail, with descriptions of their locations as fixed by cross bearings from the land, the fish species associated with each, and the times to fish there. 'It was soon obvious to us, from the spread of such grounds, that Muriwhenua fishermen had worked the whole of the inshore seas and that all workable depths were known.'

The tribunal had heard oral evidence from elders who recalled the relationship with the sea. 'Fish stories are apocryphal, in anyone's language, but those we heard in Muriwhenua were too often affirmed and corroborated too well to defy belief. They tell of native communities so bound to the sea with a wealth of laws, customs and skills, and who once enjoyed a supply of fish so bountiful, that it sometimes seemed we were in another country.'

McCully Matiu of Ngāti Kahu had described in detail a lure used to attract sharks in order to get their teeth. 'Once taken by the shark, it was played with until the teeth came out embedded in the lure. It was not Māori practice to take life needlessly. The shark's teeth would grow again.'

Wiremu Paraone recalled that his people would herd snapper up a channel, which would then be blocked off and fish taken according to need. 'Members of our community will tell you that not only is this true, but some of them even took part in this venture.'

The tribunal found numerous historical references to the way Māori engaged in trade, supplying much of the young town of Auckland. In 1852, for example, 1,792 canoes had brought 200 tons of potatoes, 1,400 baskets of onions, 1,700 baskets of maze, 1,200 baskets of peaches, 1,200 tons of firewood and 45 tons of fish.

Snapper was almost the only species of fish sold by Māori and fetched a shilling a bundle – a bundle containing anywhere from three fish to half a dozen or more, according to supply and size.

In 1992 the tribunal reported on the fishing interests of the dominant South Island tribe, Ngāi Tahu. 'Fishing in freshwater lakes, rivers and streams, and in estuaries, harbours and the sea was an important part of the economy of all iwi throughout New Zealand at the time of the signing of the treaty,' its report said. 'Fishing technology was highly developed, and apart from the selective adoption of iron barbs and hooks, and at times European fishing vessels, Māori fishing techniques had been little changed by European contact.'[7]

The oral testimony to the tribunal was notable for the beauty of language and thought. 'When the European came he bought [sic] among other things sheep, gorse, stoats, wheat, fruit trees and trout,' Rawiri Te Maire Tau was recorded as saying. 'With the Pākehā trout came his laws. They were placed in our waterways, our garden, Te Marae o Tangaroa.

For me to catch a trout, I have to pay a licence for this privilege even though it is destroying my garden.'

Ngāi Tahu did not claim they had fished in what had become the exclusive economic zone, nor that they knew its many species of fish. What they said was that under the treaty they had chiefdomship – rangatiratanga – over the waters.

Traditionally they had recognised that the waters around the South Island varied greatly. Off Kaikoura at the top of the South Island, where humpback and southern right whales gathered, shallow water quickly gave way to deeper ocean. In other places the shallow water ran much further out. They recognised the difference between demersal and pelagic fish, the former living close to the seabed and the latter in mid-water or near the surface. And knowledge of fish species was extensive: coastal pelagic kahawai, trevally, jack mackerel and barracuda; oceanic pelagics such as albacore and skipjack tuna, swordfish and striped marlin.

Ngāi Tahu leader Tipene O'Regan argued that the waters around the South Island were always dangerous so Māori had to have strong canoes capable of going to sea. '[Ngāi Tahu] had the capacity to fish where they needed to and sometimes that meant some distance from the shore. When better vessels became available they acquired them and then they built them for themselves.'

The tribunal found that fishing had formed an essential part of the Ngāi Tahu economy prior to 1840. Fishermen had been capable of going out beyond what are today territorial waters – 12 nautical miles – in good sea conditions but there had been little need for them to do so: 'Any more extensive or labour-intensive exploitation of their sea fisheries was simply not necessary.' Barracuda was their main catch, but they also took red cod, wrasses, blue cod, ling and hāpuku.

In 1986 the New Zealand government introduced a quota management system. A quota is essentially a right to catch fish – a share of the fishable stock in any species. Quotas can be owned by fishing companies, Māori iwi, and individuals, and they can be mortgaged and on-sold. Thus, if the government sets the annual allowable commercial catch of snapper at, for instance, 1,000 tonnes, and a fisherman or company has a quota for ten percent, they can take 100 tonnes of snapper. If they take more they have to pay a 'deemed' fee or penalty, which is often much higher than the value of the fish. Because of this, fishermen who end up with a catch bigger than their quota can be inclined to quietly and illegally throw back the excess fish.

The allowable catch is adjusted every so often – for example, in the 1990s the allowable catch of hoki was cut because scientists found the biomass had fallen; later it was raised as stocks of hoki recovered. Within the system, allowances have to be made for recreational fishing – especially for a species such as snapper – and for traditional indigenous catches.

When the quota management system was introduced, the promise of an expanded and lucrative industry offering high employment meant it was politically acceptable for the government to negotiate terms with Māori. There was compelling evidence that Māori had a traditional relationship with the sea, and the Treaty of Waitangi defined the fisheries as a taonga, or treasure.

Settlement with all iwi came in 1992. The combined tribes were given ten percent of the total existing quota, plus a half share of a fishing company, Sealord, and 20 percent of any future quota for fish stocks brought into the management system. Limitations were placed on the tradeability of quota that iwi received from the government: it could be sold only to other iwi or to a new body called the Treaty of Waitangi Fisheries Commission. (In 2004 this body would be rejigged as Te Ohu Kaimoana, The Māori Fisheries Trust.) The government purchased the half-share in Sealord, then owned by corporate raiders Brierley Investments, and transferred it to Māori via the commission.

On the face of it the settlement seemed an opportunity to provide Māori with large numbers of jobs, including in provincial areas where work was often hard to come by. Initially these jobs ranged from low-wage factory worker to seaman and manager. However, the work quickly became globalised. Fishing companies – both Māori-owned and others – moved to a low-wage model and the jobs went wherever the cheapest labour could be found.

Like other quota owners, Māori iwi sell their entitlement to brokers; the brokers consolidate quotas so big fishing companies can make use of them. This can be useful. A tribe in the far north of New Zealand, for example, may find itself with a quota to fish for southern blue whiting or squid near the Auckland Islands. It would be uneconomic for the iwi to do this as a solitary operation, so its quota is packaged up and sold by a broker. It will end up on a boat that is fishing for a variety of species, using quota from a number of different iwi. The boat will usually be flagged to a foreign nation.

Quota-consolidating companies can make a fortune; they are the ultimate middlemen, distant from the food chain and the ocean.

When I investigated the slave-like operations on foreign-flagged fishing boats, I found Māori unwilling to discuss their role. Hawke's Bay's Ngāti Kahungunu, the country's third largest tribal group, had just opted out of fishing their quota and passed it to a consolidator. When I asked why, an official, Aramanu Ropiha, replied: 'There are protocols of courtesy should media wish to engage with Ngāti Kahungunu. Should this iwi opt to make a statement, a written press release will be issued.'[8]

No statement followed.

It was even worse with the Māori Party. Set up in 2005 largely in protest against the Labour-led government's Foreshore and Seabed Act 2004, the party's professed aim was to fight for the rights of New Zealand's underclass. However, it appeared unconcerned about the conditions being endured by members of Asia's underclass on ships fishing for Māori quota. Its co-leader Pita Sharples, who was Māori Affairs minister in the National-led coalition government, said it would not be right for the government to interfere with how Māori managed their fishing quota. If Māori wanted to employ cheap foreign labour, he implied, they would.

Māori landowners tried 'to balance commercial, social and cultural imperatives in the way they manage their lands.'[9]

Māori's unique relationship with the sea came into question in 2009, in a case involving a small Taranaki iwi. Ngāti Tama originally came from a rugged area in northern Taranaki marked by the Mokau River mouth. During the intertribal musket wars of the early 19th century, the tribe sought safety by moving south to the area that was to become Wellington, arriving around 1820 with other Taranaki iwi, notably Ngāti Mutunga.

Ngāti Tama continues the story on its website, saying there were pressures in Wellington and the iwi 'sought opportunities further afield...' These opportunities lay 770 kilometres across the sea in the Chatham Islands. In 1835, Ngāti Tama and Ngāti Mutunga seized a ship, the *Rodney*, to get to the islands from Wellington. On arrival they attacked the inhabitants, the Moriori. In the first round, about 300 Moriori were killed and hundreds more enslaved. Some were even given as slaves to tribes on the mainland.

In the 1860s British settlers launched land wars in Taranaki, and members of Ngāti Tama who had stayed behind were caught up in the conflict. By 1868 all Ngāti Tama had left the Chathams and returned to Taranaki to fight the invaders but to no avail: the victorious Pākehā helped themselves to 27,000 hectares of the tribe's land.

In 1999, Ngāti Tama reached a settlement with the New Zealand government over this loss. It provided for a formal apology from the Crown, a payment of $14.5 million, and cultural redress, including the transfer of 1,870 hectares of land. 'I am pleased to see such progress being made toward resolving the grievances of Taranaki Māori where so much land was unjustly confiscated, so many lives were lost, and so much destruction was inflicted,' treaty negotiations minister Doug Graham said.[10] Ngāti Tama, which has around 5,000 members, had the grace not to try and assert ownership over the Chathams, where almost all Moriori had eventually been wiped out.

The treaty money came under the control of a seven-member Ngāti Tama Development Trust, of which Greg White, a prominent member of the iwi, was chief executive. In a manner he has not disclosed, in 2005 White, a former National Party election candidate, entered into a partnership with one of the many Sajo Oyang shell companies. The plan was to go into business as Tu'Ere Fishing Ltd in a 50:50 deal. Together, Oyang and Ngāti Tama would fish for tuere, otherwise known as hagfish.

Although not eels scientifically speaking, tuere are known as 'snot eels' because of their long thin bodies and the repulsive slime they use when defending themselves. They are sold only in Korea – for their flesh and their leathery skins. That there is even interest in fishing for them hints at the corrupted nature of modern fishing; the fish, which is not needed for human food and has a role in the ecosystem that is far from fully understood, is being plundered.

Sajo Oyang came up with a boat, *Shin Ji*. The declared catch was likely to make several million dollars, particularly as the company was using cheap Indonesian labour.

In 2009, when *Shin Ji* was in the port of Tauranga, its 12 Indonesian crewmen quit. The local Maritime Union called in Grahame MacLaren, inspector for the International Transport Workers' Federation. MacLaren had no doubt what was wrong. 'The vessel was in need of a good clean and there were large areas of rust on the deck in the galley, no bed linen, no hot water, with the crew expected to shower in cold sea water,' he noted in a press statement issued by the Maritime Union of New Zealand. 'The life rafts were almost inaccessible due to fishing gear being stowed all around them.'[11]

Police discovered the crew before anybody in authority on the ship knew what was happening. Due to the union's clout, the men got some of their wages and left for home.

Tu'Ere and Sajo Oyang flew in another crew.

On board ship, hagfish are kept in a tank to excrete whatever is in them; the water is flushed constantly until the eels are processed. In January

2011 the ship was back fishing when something blocked a tank holding tonnes of fish. The Korean officers sent an Indonesian boatswain into the tank; he was equipped with only shipboard breathing apparatus used in damage control and not suited to diving.

In the polite language used to describe what happened, the eels 'had access to his body'.[12] As the ship was outside the limit of New Zealand's territorial waters and flagged to Korea, no official action was taken over the man's death. New Zealand police logged it as a death at sea and the case was closed. Seven traumatised crewmen went home, leaving the remaining seven to do all the work.

Later in the year *Shin Ji* came into Auckland and this time all the crew left the ship. They managed to contact me, stating that they wanted to talk but without disclosing their location: fear of the Korean agents was strong. I spoke with the men in a beach house, with two Indonesian women interpreting. The presence of the women made some of the men's revelations awkward. As well as atrocious working and living conditions, they complained of sexual abuse. The captain would demand a massage for an hour each day, attacking them if he did not get it.

'We feel very pressured to do it and scared to resist,' one man told me. 'Every day captain wants massage for hours until after he falls asleep. If it's my turn to go on watch, it's also my turn to massage him. … feel very unhappy but no choice.'[13]

Although *Shin Ji* was placed under civil – admiralty – arrest, it quickly became plain that the ancient wreck was entirely expendable for Oyang.[14] Tu'Ere Ltd went into liquidation and the ship sat in one of Auckland's prime docks for weeks.

Eventually it slipped out of Auckland and headed to Suva, Fiji, where many of the region's wrecks see out their days under the regime of Voreqe (Frank) Bainimarama, which has become notoriously lax at enforcing environmental and shipping laws. The crew never got their money and were able to return home only with charity.

Ngāti Tama announced they had lost most of their treaty money. According to an elder, Wiremu Matuku, they had not realised what was happening. 'Our tribe is devastated, not only by the news but also by the realisation that once it becomes public Ngāti Tama will be battered by speculation over why various investments went so bad. We are going to have to carry a heavy burden of shame and accusation.'[15]

In 2001 Greg White had described himself to *The Taranaki Daily News* as an extremely stubborn man: 'If I think something is wrong I'll let people know. And if I think something is right, I'll die for the cause.'[16] This seemed to have been forgotten. He said nothing, resigned and left the scene for a time.

The Auckland University School of Business's Glenn Simmons and Christina Stringer interviewed the crew of the *Shin Ji* and in 2014 published a case study, 'New Zealand's Fisheries Management System: Forced Labour an Ignored or Overlooked Dimension'.[17] The men, they reported, were victims of forced labour. 'The treatment of the crew illustrates a number of dimensions … including deception, financial exploitation and abuse. This case highlights that if decisive action had been taken when accounts of fishing crew exploitation first emerged in the 1990s the shocking case of the *Shin Ji* could have been avoided.'

One of the crew had said that he and the other men were effectively imprisoned. 'We are slaves because normal employees have a voice, but we do not … [We] didn't expect this when we signed the contract, but once on the boat in New Zealand we are trapped…'

As with most Indonesian crews, the men had been recruited through local agents. 'We know the agent usually through friends or family, and even though we are not satisfied with the agent we have no choice as we are jobless, we can't find a job and have to be independent to help our parents, so our last choice is to work for the agent.'

Another crew member told how he was made to work all day fixing the engines. After 12 hours he tried to eat, despite the captain demanding he continue working. 'I said wait a minute to finish eating and he went crazy swearing and insulting me, calling me all the worst names, dog, monkey, pig. … I cannot take it any more.'

Simmons and Stringer also found that fatigue had produced injuries, and crew members had not been allowed to see a doctor.

When it came to being paid, the men had had to sign blank timesheets. 'Normally we work 16 to 20 hours every day,' one said. 'We have no choice but to accept whatever is written on the timesheets: if we don't sign we won't get paid. … It's pressure and force by the Koreans and the agent back home.'

After walking off the ship, the men were eventually sent home to Indonesia. When they tried to get the money they were owed, they were refused and told they were now blacklisted.

Simmons and Stringer noted in their report that low wages and poor working conditions were, by themselves, not symptomatic of forced labour. However, the withholding of pay, incarceration, retention of identity documents, and deprivation of food were. 'While the exploitation of their labour took place at sea, the trajectory to forced labour began in Indonesia, with their manning agent commencing financial exploitation,' they stated.

Why had the crew put up with the appalling conditions? One had explained: 'We think about family and the certificates they [manning agents] have back home.'

A little known aspect of Britain's abolition of slavery in 1834 was that the government paid 46,000 people, including many of the country's wealthiest and largest companies and banks, £20 million (£16 billion in today's money) to compensate them for the loss of their 'property'. A significant number of Britain's stately homes were built with this money. Slaves got no compensation for the loss of liberty and dignity they had endured.

Under the Treaty of Waitangi, iwi are entitled to compensation for changes in government policy that are contemporary breaches of the principles of the treaty. In 2012, New Zealand's newly minted Ministry for Primary Industries claimed that if iwi lost access to foreign-chartered fishing boats they would be entitled to up to $300 million in compensation.

Legislation requiring the reflagging of all such vessels to New Zealand – meaning the boats would have to obey New Zealand law, including paying crew at least the New Zealand minimum wage – might 'disproportionately impact on Māori and iwi quota holders, particularly if [other] vessels are unavailable'.

Iwi, the ministry claimed, would have to be compensated for the loss in value of some fisheries that would be unviable without cheap crews, including those of squid, jack mackerel, southern blue whiting and hake. A 'worst case scenario could result in a loss in export revenues of around $300 million annually'.[18]

10 Pirate ship

Losing a ship and six men did not mean much to Sajo Oyang Corporation. It quickly found another ship, *Oyang 75*, and crew. The factory manager was Tae Won Jo, a survivor from *Oyang 70*. It was Jo, as factory manager on that doomed vessel, who had told Captain Shin before the ship sank that he was ready for more fish to come aboard, advice that had proved so disastrous.

Sajo Oyang hired a public relations man, Glenn Inwood, to convince the public and media that all was fine on the high seas. Inwood, born in 1968, claims extensive experience in the New Zealand media as a reporter but I have never come across him in any significant role. At some point he joined the stable of hired hacks who service politicians in parliament, becoming press secretary for an immigration minister. For reasons he has never explained he became a big fan of whaling. When Prime Minister Helen Clark pointed out his 'distasteful' connections he resigned.

He then created Omeka Public Relations and picked up clients no one else would touch, including the vaguely defined Institute of Cetacean Research, Imperial Tobacco New Zealand, Japan Fisheries Agency, Japan Whaling Association, Species Management Specialists, and the World Council of Whalers. He also claims Māori connections; his Omeka website describes him as 'affiliating to Ngāti Kahungunu and Tūhoe iwi'.

In Inwood's opinion, the Japanese have the right to hunt whales. 'That's not to say that I don't think whales are magnificent creatures. I just don't believe they are sacrosanct...' Paul Watson of Sea Shepherd, an international group fighting Japanese whaling and other slaughter of oceanic wildlife, regards the PR man as a 'priceless asset': 'A good public relations man should at least understand his opposition but Inwood does not. He should also understand the basics of diplomacy but Inwood does not.'[1]

I had already clashed with Glenn Inwood over Sajo Oyang's operations, and he had responded to some of my questions with a drawing of a dog turd. He then launched a bid to find out how I was getting information on the fishing industry, filing requests under the Official Information Act with the University of Auckland and various government departments. He demanded copies of emails, letters and other forms of communication the university had had with me and others, and all the documents held by the business school on foreign charter vessels.

When that got nowhere, he wrote to *The Sunday Star-Times* claiming that Sajo Oyang, its officers and crew and representatives, had never been the subject of a prosecution. He was wrong but that was not surprising: he and his clients believe there is a left-wing conspiracy to destroy their businesses. He objected to my describing Southern Storm Fishing Ltd as an Oyang shell company, saying the media were incapable of distinguishing 'between separate entities or agents and who is acting for whom.'[2]

Inwood's foray into Korean fishing would become a personal embarrassment for him. The *Oyang 75* saga was destined to finally end the cosy relationship the fishing industry had with the New Zealand government and its regulators. The revelations that emerged would resound both legally and politically.

In May 2011, when *Oyang 75* arrived in Dunedin, Inwood invited media to come aboard and see the boat as a model of modern fishing. State-owned Television New Zealand accepted the bait – and Inwood's assurances that Sajo Oyang was moving past the stories of human rights abuses.

The Ministry of Fisheries, which had assiduously ignored human rights abuses on foreign-chartered fishing boats, placed four officials on board *Oyang 75*. It claimed this was routine involvement with a new ship in the region. Inwood had another spin: the officials were there to 'upskill, to be more effective managers of New Zealand's world-class quota management system … a rare accomplishment in an industry as highly regulated as the New Zealand fishing industry and speaks volumes as to the level of commitment that both companies have to compliance with New Zealand fisheries laws.'[3]

Twenty-two-year-old *Oyang 75* was an unusual ship for a Korean company. Built in Spain, it was equipped with better crew accommodation than were Sajo Oyang's usual vessels. But like much else in the world of international fishing, its name was a charade. It had originally been called *Garoya Segundo* and owned by a Spanish company, Grupo Oya Pérez, well known in fishing circles for illegal, unreported and unregulated fishing in the Atlantic. Between 2000 and 2005 one of Grupo Oya Pérez's fishing boats, the *Ross*, had managed to become one of the world's worst perpetrators of illegal fishing, with seven arrests in the northwest Atlantic. It had then sailed to the Southern Ocean to illegally loot toothfish.

In November 2005 *Garoya Segundo* had been caught illegally fishing in Norway, taking Greenland halibut in a protection zone in the Barents Sea near Svalbard. When caught, the ship had on board 508 tonnes of halibut – 308 tonnes more than it was allowed under a research quota granted by the Spanish government and worth one million euros. The captain had manipulated daily catch reports and provided incorrect reports to Spain's fisheries ministry. After diplomatic protests and payment of a fine of US$1.8 million, *Garoya Segundo* was released.[4]

As is common when ships are exposed for illegal activity, *Garoya Segundo* underwent name and flag changes. Renamed *Amor Saco*, it was sold to Sajo Oyang and then headed to New Zealand as *Oyang 75*.

After Glenn Inwood's public relations exercise in Dunedin, *Oyang 75* put to sea and made two fishing trips. Each was scarred by abuse and assaults by the Korean officers towards the Indonesian crew. The day after the boat sailed into Lyttelton on June 18, 2011 at the end of the second fishing trip, police, alerted by Christchurch Hospital emergency staff, visited to investigate a complaint of assault on a crew member by the ship's chief engineer.

As had become the pattern when New Zealand police investigated incidents on foreign boats, no further action was taken. The crew would not accept this and all 32 Indonesian workers walked off. The Southern Storm agent Pete Dawson first threatened the men and then made efforts to get them out of the country quickly. This would have worked in the past – under New Zealand's immigration system, if an agent says a foreigner is no longer employed the person's visa expires and they have to leave. But those of us working away at the issue were starting to get political leverage. The Ministry of Fisheries and the Department of Labour had been exposed as little more than rubber-stampers for the industry. The publicity was hurting public servants and things were about to change.

Even Oyang knew it had to be careful. The company's chief executive, Il-Sik Kim, sent emails to Southern Storm and letters to their agents in Jakarta over the fate of *Oyang 75*'s crew. 'Oyang strongly desires that you and other crew agent companies do not take steps to subject these ex crew members to any type of penalty on their return home,' Kim warned. 'Please be on notice that if crew agents do take steps, this will have serious implications for Sajo Oyang's business in New Zealand and may force Sajo Oyang to look for alternative sources of crew in the future.'⁵

The company had learned that, while most of the crew had been moved out of New Zealand as Dawson wanted, six had been quietly diverted at Christchurch Airport to tell authorities what was really happening on the Korean boats. Sixteen other crew members had made statements when interviewed ahead of their departure. A judge would late rule the men had been 'honest and reliable' in their accounts.

The evidence showed that *Oyang 75* was nothing more than a pirate ship. In June 2012 the New Zealand government charged five of its officers with dumping fish, making false or misleading statements, and hindering a fisheries observer.

The master, Chong Pil Yun, faced ten charges, including three that he jointly offended with others in aiding and abetting Southern Storm Fishing to return or abandon fish subject to quota. Other charges related to false and misleading statements, and a single charge related to hindering fisheries observer Elizabeth Dyer from transmitting a message to her office in Wellington. Chief officer Min Su Park faced four charges, and factory manager and *Oyang 70* survivor Tae Won Jo three charges. There were two charges against the bosun, Wongeun Kang, and two against the radio operator, Juncheol Lee, over false returns. None of the men would show up for the trial.

The case, when it came before the court, offered a vivid picture of how Korean foreign charter vessels were ripping off New Zealand.

Elizabeth Dyer had boarded *Oyang 75* as it departed from Port Chalmers, Dunedin, on March 8, 2011. At sea she had tried to tell the Korean officers how to use a system for counting discarded fish as they arrived on the factory deck. They had ignored her. She and another observer worked in shifts to try and stop the ship dumping hoki. Tae Won Jo yelled at them to get out and crew were ordered to continue discarding fish. Dyer had stopped pointing out to the officers they were doing something illegal: she had not wanted to upset and antagonise them.

The practice the ship was engaging in is known as 'high-grading' – dumping low-value fish and replacing it with fish of higher value. Ships cannot legally do this: they are required to take quota catch as it comes and process it through to freezing, but the Koreans were thinning out the smaller and damaged fish.

Hoki, which can be easily damaged if left in nets, cannot be processed later. The observers on *Oyang 75* were seeing fish on the factory line and then observing boxes going into the freezers. These fish, which the men called 'bulong', were to be thrown overboard. While the observers were asleep, a human chain would be formed to get the boxes out of the freezer and over the side. On its first voyage *Oyang 75* dumped fish worth around NZ$755,000 landed at the wharf; on its second voyage it dumped fish worth around $1.4 million.

Dyer told the court it was apparent the large catches being taken meant the factory could not keep up with the processing. She noted that the crew worked excessively long shifts when fish were being caught. At one point a crewman had even fallen asleep at the fish pound, where fish were held before processing. She had tried to fax details of the operation to her office but the captain had snatched the paper out of her hand. He later denied doing this but the court found that it was unlikely Dyer would fabricate evidence.

While on board, Dyer had also noticed that Park shouted at the crew, who cowered away from him, and had observed Kang hit a crewman on the back of the head.

A second *Oyang 75* trip began on May 15, 2011. This time Dyer was joined on board by another observer, Susannah Barham, who found the experience gruelling. She told the court of being summoned to dinner in the officers' mess one evening and finding the captain, radio operator and bosun drinking soju, a vodka-like alcoholic drink distilled from rice. By the end of dinner both the captain and the bosun were drunk. The bosun followed her to the factory area. 'He had been entertaining himself by locking a crew member in the fish pound. I was appalled by his behaviour and intervened because I saw it as a safety issue,' Barham told the court.[6] Because the bosun continued to follow her around, she went to the bridge to be in the presence of other people.

She noted other senior officers also acting in an aggressive way towards the crew. Kang gave crew 'a quick smack across the head' and threw fish at them. Jo and Yun yelled at the men in angry aggressive tones.

As on the previous trip, crew would work up to ten hours to get processing completed when a trawl had been carried out, and there would be only a few hours' rest between trawls. There was no shift system operating.

Thirty-seven-year-old Slamet Raharjo, an Indonesian man who had remained in New Zealand to give evidence for the state, reported that *Oyang 75*'s crew worked continuously when the vessel was fishing, with only short breaks for eating. The longest of these work periods was two days and two nights.

He had never seen so much fish dumping in his 17 years as a deckhand. A fellow deckhand, Aroyo Aristiary, also told the court that significant amounts of fish had been thrown overboard. Two other crew members testified to seeing the captain inspect the pound and order fish to be discarded.

Judge David Saunders read out an affidavit from one of the Indonesian crew, in which the man said he had been a deckhand for two to three hours each day and then had to work in the factory. His average working day was 18 hours. On his first trip about 30 percent of the total hoki catch and half the squid catch were thrown overboard. On the second trip, crew were told not to throw away the hoki while they were being watched. It was to be done when the observers were not around.

The judge was blunt. The clear theme of the evidence he had heard was that the Koreans were out to maximise their return on each trawl. 'The repeated reference to a "Let go, let go" instruction was such that one could only conclude the focus was on seeing that any inferior species or damaged fish not be recorded or taken into account as quota species.

'It is apparent that the Indonesian seamen were, in the main, ignorant of the fisheries regulations and were simply not in a position to countermand the instructions given by the Koreans…'[7]

The fact the Korean officers had not shown up for the hearing showed they would not allow themselves to be held accountable. They had been demanding, arrogant and high-handed towards the crew – and towards the law itself. 'The unchallenged evidence effectively leaves the court with the distinct impression that the Korean officers ruled the crew with an iron fist and were engaged in practices which, by any measure, would not, and should not, be tolerated in any civilised work environment.'

He fined the four officers more than $420,000 in total and ordered forfeiture of the ship, worth up to US$8 million. Sajo Oyang appealed the forfeiture; this was still before the courts as this book went to press.

While a New Zealand court had had no problem finding conditions aboard the fishing boat dreadful, in Korea the matter was greeted with indifference. *Oyang 70*'s sinking had been hardly covered at the time and the official safety report out of Seoul had been perfunctory. Now, with *Oyang 75*, the National Human Rights Commission of Korea investigated only after receiving complaints from three bodies – Korean House for International Solidarity, Advocates for Public Interest Law and Center for Good Corporations – that the ship's Indonesian sailors had suffered physical and verbal violence, sexual harassment and wage discrimination.

The commission dismissed the complaint about physical and verbal violence, saying the issue fell 'within the category of human rights violation between private persons, which does not belong to the commission's scope of the matters subject to investigation'. It said it was not possible to discern pay discrimination 'since they are in different positions with different responsibilities', while an allegation of sexual harassment was dismissed due to lack of objective evidence.[8]

The crew of *Oyang 75* had little success at getting paid the pittance they were owed. The accountancy firm PricewaterhouseCoopers prepared a report for the New Zealand Department of Labour in which it tried to work out where the crew wages from *Oyang 75* had ended up. I obtained a copy through the Official Information Act. It defined the relationships between the various parties involved. Sajo Oyang Corporation owned the ship and was the employer of the Indonesians. The shell company Southern Storm Fishing leased the quota and had hired Sajo Oyang to fish it. The crewing agents had sought out 'employment opportunities on behalf of individual crewmen' and 'dealt directly with the ship owner'.[9] The agents, all Indonesian, were named as Pt Nurindo Mandiri International, Pt Panca Karsa Mandiri Sejati and Pt Oriza Sativa Agency.

PricewaterhouseCoopers staff had tracked the story of the six crewmen. Between them, the men had come from all three agencies. Each had had to pay an upfront fee, varying from 2.5 million rupiah (around US$215) to 3.8 million rupiah (US$330). Each had signed a contract but none had been given copies. Each had an expectation of an agreed monthly wage, ranging from US$220 to $480. Only two had received their agreed payments; the others had lost part of their pay to the agents. The gaps between what they had received and what they were supposed to receive was not explained. 'The crewing agents have refused to provide any information which might explain the differences and which might assist to account for the sums involved,' PricewaterhouseCoopers stated.

Legal authorities in Korea had a look at the report. The Busan regional public prosecutor, Yoo Kyungpil, then filed his own, saying an unnamed CEO of Sajo Industries had admitted to his 'suspected crimes' around the crew's pay. It was the first time the man had been accused of criminal activity; he would not be prosecuted.

While the crew had stated they were owed a total of US$1 million, Sajo Oyang claimed there was only $260,000 unpaid. Sajo Industries was 'not in a situation' where it could pay that immediately, the prosecutor said. It had held an emergency meeting on the morning of November 27, 2011 and decided that, in order to be able to negotiate with the New Zealand Immigration Office, it needed to forge foreign currency transfer receipts of each of the 32 crew members 'in the exact same format as their usual bank … issues to make it look like they had paid the $260,000'.[10] An official at Sajo Oyang had erased the transfer date, number and amount on a previous receipt and printed it as a blank form. He had then made 32 copies. When the company discovered it did not have employment contracts in Korea for the men, another official had forged them and faked the signatures.

In July 2011 Sajo executive Myeong-Ho Lee, visiting New Zealand to try and resolve the deepening mess his company was in over *Oyang 75*, revealed in a meeting with the company's local connections the modus operandi of Korean fishing companies in New Zealand waters. 'If we paid like the Kiwis our business cannot align, this thing must be finished. So that's why we have to invite Indonesian crew, why we didn't invite Korean crew: the difference is the salary.'[11]

Paying New Zealand's legally sanctioned minimum wage was impossible, he claimed. 'If we pay the same as the Kiwis, no reason to do business here. That's true so that's why position has never changed – same as before. If we pay the full sum … this business has to stop.'

He asked about the New Zealand minimum wage and was told it was $14.80 an hour. His accountant's mind kicked in. 'I briefly calculate it, okay? Thirty-two crew – okay 30 crew – one day $225, this one times 30 days, $6,000, so that why I cannot pay the Indonesian crew, we have no money, no business any more. Calculation 6,000 × 30 persons × 365 days, 12 months, how much, times 12 is $2 million. Two million dollars for crew additionally; in this case our business is gone, minus.'

Lee was pleased with the way his and other Korean companies had come up with ways of deducting costs from the crews' pay, especially for things such as clothing. 'This one made by clever lawyer, clever lawyer, lawyer made, not real one.'

There was a system of double accounting to control both the payments and who could see what money was going where. 'Each crew account is made by their own signature, by the power of attorney, but this account is [actually] agent's – falsified. Forget about that, okay. That's very political. It's the lawyers – they control this account.'

Lee's cynicism came to the fore even more when he talked about applying to Immigration New Zealand for an 'approval in principle' – that is, conditional approval to hire workers from overseas. 'When [we] apply [for] the AIP we must say we guarantee to do the rule of New Zealand law, blah blah blah blah – so that's why we can bring the crew here. And then you explain that – all the expense – you can't afford to pay New Zealand wages, right, because you spend too much. And even [though] you talk talk too much to immigration, they will not understand you, they will not agree. … The lawyer who concerned with labour, he said that immigration service doesn't interested about the law, they really doesn't understand the law.'

In one meeting he described chairing a secret committee in South Korea that controlled the whole Korean fishing operation in the South Pacific. As well as Sajo Oyang, the committee had representatives of other companies operating in New Zealand, including Dong Won Fisheries Co., which operates with Auckland fishing company Sanford. 'They talk and discuss every day … Dong Won is ready to pay fishing bonus up every month,' Lee said. 'When I return to Korea it will be decided.'

Such collusion is against New Zealand law.

The Sajo boss knew that sending the *Oyang 75* crew home would cost him money. 'If they return to Jakarta the Indonesian agent will raise the suit, claim damages, sure, 100 percent, so they lost big money and risk to jail. They thought [we] going to pay them every month more money. Every man? Every month? Oh, every month, ha ha.

'We no ready to pay. Even we lost a million dollar we are not, no never. … Inhumane treatment: my fault, our fault. Poor crew. My final offer, because now everything is the lawyer's decision. So I cannot control every present situation. The ball is not my court – gone.'

He had good reason to hold out. Ultimately, he knew, Indonesians did not want to be blacklisted by Korean fishing companies.

There was another interesting revelation at the meeting: Lee divulged that Sajo had a private detective investigating Glenn Simmons and Christina Stringer. He named the detective. Oddly I knew him in a different context – our daughters had attended the same school.

The scandal of *Oyang 75* continued. After the publicity that Simmons, Stringer and I had generated around the human rights abuses, various other branches of government began to take an interest. Maritime New Zealand found the ship had hidden plumbing and a secret switch hidden under the floor plates of the engine room so it could discharge unfiltered bilge, including oil, straight into the sea unnoticed. There was evidence the piping had been used at least twice. There was a legal requirement to notify the agency if harmful substances were discharged or escaped into the sea, but there had been no such notification.

Under the International Convention for the Prevention of Pollution from Ships, vessels must use an oily water separator to remove harmful substances from any wastewater going into the sea. Maritime New Zealand's Paul Fantham said the rules were clear and the organisation could not allow operators to damage New Zealand's marine environment. In February 2013 the Christchurch District Court fined Southern Storm Fishing $10,500 for the violation. This was the company whose actions, according to Glenn Inwood, spoke volumes about its commitment to New Zealand law.

Sajo Oyang got *Oyang 75* out of New Zealand. It began operating out of Port Louis in Mauritius, where Greenpeace would accuse it of plundering vulnerable fish stocks.

Meanwhile the ship remained under a bond to the Ministry for Primary Industries. In theory the money will be forfeited if *Oyang 75* does not return to New Zealand but the government may not press the company. In July 2013 the prime minister, John Key, paid an official visit to South Korea. He was anxious to come up with a free trade deal. Seoul has made it clear that, in return for free and fair trade, New Zealand will have to give South Korean vessels free access to its exclusive economic zone.[12]

Early in 2014 the Korean Coast Guard Police laid charges of fraud against four Sajo Oyang officials – including the fishing vessels manager, Gap Suk Lee, and the administrator of trawling operations, Geon Taek Kim – for forging foreign currency transfer receipts indicating the Indonesian crew of *Oyang 75* had been paid. The men were convicted in April 2014.

Further charges were pending against five officers on *Oyang 75* for crimes ranging from assault to sexual violence.

11   The politics of fishing

While inshore fishing in New Zealand goes back to the arrival of the Polynesians, the country's geographic isolation means industrial-scale fishing has happened only in recent times. Deep-water commercial fishing began in the early 1960s. The 1977 declaration of a 200-mile exclusive economic zone was followed in 1982 by the United Nations Convention on the Law of the Sea, which confirmed New Zealand's full sovereignty over a smaller, 12-mile territorial limit and the right to manage its exclusive economic zone – which for simplicity I'll term the 'deep water'. (In reality not all deep water is in the zone; nor does water have to be deep to be in the zone.)

There was a caveat: although New Zealand got to manage the deep-water fishery, if the country was unable to catch enough fish, or kept out other genuine fishers for no valid reason, other nations could claim fishing rights.

By the late 1970s, New Zealand still had only limited capacity to fish in the deep water. Catch that was beyond the capacity of domestic vessels was allocated to foreign countries under government-to-government licence agreements. Ships from Japan, Korea and the Soviet Union were permitted to fish in New Zealand waters and return their catch to foreign ports for processing.

These vessels were catching in excess of 200,000 tonnes of fish each year. All this was expected to be temporary as New Zealand built up its own skills and fleet. However, science changed the game: when new fish species were found in the zone, they offered the prospect of big money but the costs of extraction were high.

The Chatham Rise is a vast ridge extending eastward from southern New Zealand a thousand kilometres to the Chatham Islands. Look at it on a chart and it resembles a nautical alp, with hills, ridges, peaks and volcanoes. The water above the ridge is only ever about 1,000 metres deep, but north and south of it the ocean falls away to around 3,000 metres.

Only a name to most New Zealanders, the Chatham Rise is among the country's most lucrative real estate: it is here that the prized catch of orange roughy come in large numbers to spawn. In the 1970s, commercial trawlers from the Soviet Union first discovered these spawning aggregations. After New Zealand declared its exclusive economic zone, local boats hit the Rise. In 1979 they took around 54,000 tonnes of orange roughy, with vessels reporting catches of 41 tonnes a day. Twenty or so large factory trawlers, dominated by those of Sanford, a large Auckland-based company, hit the flat areas and were responsible for about 60 percent of the catch.

Inevitably, and quite rapidly, the amount of fish caught started to fall as the seamount fisheries were plundered: a rough guess at the total weight of orange roughy taken has put it at as much as 589,000 tonnes. Given the scientific best guess that no more than two percent, or around 11,000 tonnes, of orange roughy could be sustainably taken each year, New Zealand fishing boats were engaged in a mad and destructive stripping of an ancient fish stock.

In 1997 another big find of orange roughy, and another plunder, took place on the South Tasman Rise, an undersea ridge extending south from Tasmania. A fishery centred on a small number of pinnacles was developed and fished by Australian and New Zealand vessels, most of it in international waters. In the first year the boats took 2,000 tonnes. In 1999, after the Australian fleet took 1,700 tonnes, the Australian government tried to close the fishery. With no law to control fishing in international waters, New Zealand boats took another 1,600 tonnes. Although orange roughy was also found in the eastern Atlantic and off South Africa, it was the New Zealand and Australian stocks that created a worldwide market.

Word quickly spread. Three South African vessels and a Belize-registered vessel began operations. It is estimated that between them they took 4,650 tonnes of orange roughy, effectively destroying the stock. When there are weak or non-existent regulations, some fishing companies can be counted on to go after the very last fish.

In the 19th century, neither Māori nor early European settlers were aware of orange roughy. Even today little is known about this mysterious creature of the deep. When it is fished up it is almost always dead, ruling out any kind of tag and release programme that could increase scientific knowledge.

Orange roughy produces a white, boneless, mild-tasting fillet that is amenable to freezing, giving it high value. From autopsies, scientists have learned that the species is slow to grow and slow to mature. A member of the ingloriously named slimehead family and given its present name by the United States National Marine Fisheries Service, it lives up to 149 years, and spends that time in waters between 180 and 1,800 metres deep. It is considered 'unproductive', meaning that only between one and two percent of its estimated biomass, or population, can be sustainably fished each year.

Feeders on zooplankton, as well as smaller fish and squid, orange roughy are given to forming large aggregations around seamounts and canyons where water movement is high; because of this they are attractive feed for sharks, cutthroat eels, merluccid hakes and snake mackerels. Sharks of which little is known are a common bycatch and most are discarded, even though one, the longnose velvet dogfish, is 60 to 70 percent liver oil by weight.

Just as *Tai Ching 21* was able to track tuna in Kiribati waters thanks to modern technology, it was the creation of deep trawling techniques, together with increased knowledge of seamounts, that made it possible to harvest this once unknown fish.

Orange roughy is caught by demersal trawling, in which a large net is towed just above the ocean floor. Although the practice is not as environmentally damaging as bottom trawling, studies before and after orange roughy fishing show major harm.

Around 70 percent of the catch takes place on seamounts – cone-shaped, mostly dormant, underwater volcanoes that rise steeply from the seabed. Often large, these seamounts contain great biological diversity and are vulnerable to disturbance. The most obvious is damage to corals. As the Woods Hole Oceanographic Institution has noted: 'Deep-sea corals that thrive on and around seamounts host more than 1,300 different species of animals; some [of these animals] are unique to seamounts themselves and some live only on a specific species of coral.'[1] In some cases, on seamounts subject to orange roughy fishing, there is today just bare rock where there was previously a wide diversity of life.

By 2012, New Zealand had 100 species or species groupings subject to its quota management system. Of these, 38 were being fished in the deep water, with 13 species mainly targeted. While almost all of the deep-water-caught fish are exported, the country is a small player in the world fishing economy, accounting for just 0.5 percent by volume of wild fisheries production across the globe. Wild fish exports are, however, New Zealand's fourth highest export in value; in 2011 they were worth $650 million.

Seventeen companies deep-water fish, led by Sanford out of Auckland and Sealord and Talley's out of Nelson. In 2011 the fleet comprised 56 vessels, 27 of them foreign-chartered.

The bigger fishing companies have alliances with major off-shore fishing companies. Sealord provides an example of the complex corporate structures involved. In theory it is owned by Kura Ltd, an obscure company that is, in turn, 50 percent owned by Aotearoa Fisheries Ltd. Aotearoa's shareholders are Māoridom's 57 iwi and Te Ohu Kaimoana. The other half of Sealord is owned by Japanese company Nippon Suisan Kaisha Ltd, more commonly known as Nissui. This gives Sealord access to Nissui's global network of subsidiary companies, affiliates and partners.

The relationship shows the kind of people with whom fishing companies can end up working. For a long time after it was formed, Sealord could not be certified 'dolphin safe' because Nissui was selling canned whale meat and owned 32 percent of Kyodo Senpaku Kaisha Ltd, which carries out Japanese whaling under contract to the so-called Institute of Cetacean Research, the organisation that hired Glenn Inwood to spin its pro-whaling message. Eventually Nissui bailed out as a result of international consumer pressure.

Despite Sealord being one of the dominant players in New Zealand fishing, the record shows it is not very well managed. The company has been in heavy seas for a while. In 2011 it had to borrow $69.9 million from Nissui to keep going. Then in 2013 it had to walk away from a disastrous hoki, toothfish, whiting and warehou venture off Argentina's Tierra del Fuego.

Sealord had purchased Argentina's failing Yuken SA in 2009 for an undisclosed price. Shareholders were told the new operation would turn over $60 million within five years, but it made only losses – and was funded, in part, by cutting 350 Sealord jobs in the South Island and selling mussel farms.

Not all of Yuken's failure could be blamed on Sealord. Soon after Sealord purchased the company, Argentina's economy collapsed and the country went into a period of 30 percent annual inflation and political and social instability. On top of this, Yuken's main target, hoki, was an unproven and little understood resource in that part of the world. It was open to question why a company would invest so much in something that science had yet to demonstrate was viable.

The chair of Aotearoa Fisheries Ltd, Whaimutu Dewes, was reported as saying that Sealord's loss in Argentina had had a heavy impact on the company's 50 percent Māori ownership. 'Our shareholders are unhappy … They hold us accountable in a very intimate and very direct way. … It is an extraordinarily bad and disappointing result.'[2]

Another leading player, Sanford, also has complex links. The company and Korea's Dong Won Fisheries Co. Ltd each own a 50 percent share in China's Weihai Dong Won Food Company. In the world of business and the eyes of the New Zealand government, Weihai Dong Won is a triumph, a seafood plant adding value to New Zealand produce. In reality, the company has stolen jobs from New Zealand. Once upon a time, shiploads of beheaded, gutted and snap-frozen fish would come ashore in New Zealand to be further processed. Now they go directly to China, where hundreds of low-paid women do the work. Fish are born with small bones, which customers require removed. Outfits such as Weihai hire women to remove these bones by hand.

There is, though, an economic trade-off. The fact the fish is thawed in China and then refrozen lowers its value. The system is designed to cut labour costs, not improve quality. More onshore processing would result in a greater overall economic return to New Zealand through increased employment opportunities, but mean lower overall profits for fishing companies.

The business of getting fish out of New Zealand with the barest minimum of local input is further influenced by the nature of world trade. A New Zealand-flagged boat cannot send its catch to Korea or Japan without paying a hefty tariff – for example, 22 percent on squid to Korea. A New Zealand vessel that catches 1,000 tonnes of squid and exports it to Korea would incur a tariff of NZ$369,600. Catch the same fish from a Korean boat, even under charter to a New Zealand company, and the tariff magically disappears. The squid, even though caught in New Zealand and able to be labelled 'Produce of New Zealand', is officially considered a product of Korea.

Fishing has increasingly become a political issue, so it is perhaps not surprising that fishing companies give generous campaign funding.

Electoral Commission returns show that at the 2011 elections Clayton Cosgrove, a Labour member of parliament, accepted $17,500 from Independent Fisheries of Christchurch. Some of this company's catch came from *Oyang 70* and other Korean boats used by Southern Storm Fishing.

Interviewing politicians about the source of their campaign funding is a difficult business. Few like being challenged, especially in an area attracting headlines. Initially Cosgrove told me he had had no difficulty accepting the money from Independent Fisheries, saying the company was based in his electorate. In fact it wasn't. As the former Waimakariri MP, he had, however, drafted a bill to ease restrictions on residential developments in an area that included land owned by Independent Fisheries.

He went on to tell *The Dominion Post* that Independent Fisheries' chief executive, Mike Dormer, was an old friend. 'I know his family, he knows mine, he's one of the biggest philanthropists in Christchurch, a highly respected businessman. I'm proud to say I'm a mate of his.'[3]

'Have [Independent Fisheries] been promised anything from me as an Opposition MP?' he said when I interviewed him. 'No, they have not. They have chosen to support me and that is the stone-cold end of it.'[4]

He denied overlooking the labour abuses aboard foreign charter fishing boats, and said he fought for workers' rights – 'Always have, always will, a life member of the Labour Party, been in since I was 14. I don't have to justify my credentials to anybody.'

Another Labour MP, Shane Jones, in 2011 received $10,000 from Sealord. The company's chief executive, Graham Stuart, said the industry had been well served by fisheries ministers, and noted that Shane Jones was a former chair of Sealord's board of directors. Sealord believed having someone with his hands-on experience and deep knowledge of their industry in parliament was 'beneficial. We did not make any other donations'.

Jones pointed out that he had received the same amount from Talley's in 2008. The two companies' positions were 'polar opposites'. There was no quid pro quo, he told me. Sealord acted properly with respect to the crews on its foreign charter vessels and was critical of the Korean boats. 'They are egregious, those instances that are related to the Korean fishing interests. Unfortunately, they have coloured everything in a negative light.'[5]

There were plenty of other recipients of donations from fishing companies in 2011. Talley's gave $5,000 apiece to National MPs Eric Roy, Colin King, Chris Tremain, Joanne Goodhew, Todd McClay, Lindsay Tisch, Chris Auchinvole and Chester Borrows. United Fisheries, a Christchurch company that has used foreign charter vessels later shown to have underpaid their crews, gave $3,000 to Labour candidate Megan Woods and $2,800 to National candidate Nicky Wagner. Both were elected. Woods said she had accepted the money after visiting United Fisheries, which was a large employer in her electorate of Wigram. She had not discussed with the company its use of cheap Third World fishing-boat crews.

12 Aboard a Soviet trawler

One of the enduring movies I remember from childhood is the 1954 classic *The Caine Mutiny*, in which Humphrey Bogart plays Lieutenant Commander Queeg, the paranoid captain of the destroyer-minesweeper USS *Caine*. In moments of tension, Queeg would pull two ball bearings out of his pocket and click them in the palm of his hand. The greater the tension the more he worked the bearings; it became an enduring piece of symbolism for pressure on the high seas.

When, in 2012, I found myself aboard the 4,407-gross-tonne factory trawler *Aleksandr Buryachenko*, life imitated art as the 49-year-old captain, Yuri Kylybov, paced the bridge and clicked away on a small ring of prayer beads. I didn't know a word of Ukrainian but it was easy to tell when things were not going well. Kylybov later told me that the pressures of being both a fisherman and a mariner doubled the stress – he not only had to keep his ship and crew safe, he had to fill the ship with fish.

As my coverage of foreign charter vessels continued and data emerged from the Auckland University Business School, there was growing concern from a variety of political and industry groups. The public was beginning to realise that parts of New Zealand's fishing industry was, if not corrupt, at least in the hands of foreigners who not only plundered the resource but brutalised the crews. Occasional threats, both legal and physical, came my way. Assorted people let me know there was far too much money involved for it to be lost as a result of reports by a newspaper reporter and a couple of academics.

For the companies that used foreign charter vessels, the exposure threatened their way of doing business. Among these companies was Sealord. Sealord has several Ukrainian-flagged, ex-Soviet trawlers in its fleet. The boats are large, powerful – and, with Ukrainian crews, cheap to operate. Sealord's response to the controversy was to announce they would from then on have observers on all their chartered ships. They added that the media were welcome to come along.

On behalf of *The Sunday Star-Times*, I asked to go on a voyage. Sealord was not keen. Its Auckland-based communications manager Alison Sykora insisted on checking me out. We met at a café and it was plain she had vetted me with media and public relations sources. Judging from her later comments, particularly that she had had to persuade the company to let me aboard, there were concerns about what I would write. But she also got the message from my employer, Fairfax Media, that I was a serious and credible journalist. Banning me would put Sealord in the company of people like Fiji's Bainimarama, who had also banned me. Sealord was not enthusiastic but it accepted that, having put out an open invitation, it could not veto the guests.

On a warm day in January 2012 I flew to Nelson to join the 14-year-old *Aleksandr Buryachenko*. The tough seaworthy-looking vessel turned out to be moored next to *Sparta*, a Russian-flagged toothfish boat that had a couple of months earlier narrowly missed disaster in the Southern Ocean. A couple of weary, uncommunicative Indonesians were chipping away at plating on *Sparta's* shabby and damaged hull.

On the way to the wharf, Sealord executives told me I would have free access to the ship and its crew, who were all Ukrainian, mostly from Sevastopol on the Black Sea. The only thing I was not to mention in my reports was how the crew were paid. Every so often an armoured van would come down to the ship and give them their money in United

States dollars. As the rules have since been changed and government regulations now require payments be deposited in independent bank accounts, I do not mind mentioning this.

I met with Kylybov in his cabin, a large room with chilli plants growing in the portholes. There was a strongroom behind him. He said, smiling, that in the Soviet days that was where his secret coded instructions in the event of the outbreak of World War III would have been stored. *Aleksandr Buryachenko* had been launched in the Soviet era, so it had gun platforms and space for electronic surveillance equipment. Even a casual glance at its spacious bridge, far bigger than needed on the average fishing boat, suggested it was designed for much more than catching fish. The boat's value now lay in its powerful engines, which allowed it to trawl at speed and catch fish that might otherwise be elusive.

Kylybov resented any comparison between his ship and the other foreign charter vessels operating out of New Zealand. 'I can only speak for Ukrainian vessels and we have good discipline, good crews, good food and we have common culture and common values,' he said through an interpreter. 'Same country, same nation, same values.'

He felt sorry for crews that included people of different religions, nationalities and culture. 'They can end up fighting over simple things like food. … They don't speak each other's languages at all; they have only a couple of common words between crews and officers.'

After coffee and some awful cakes, I was taken to my cabin. Usually occupied by an observer, it was a veritable suite, complete with large square portholes and an entry out on to the deck on the port side of the ship. The bed was an appalling lump. A large iron bathtub dominated the bathroom. I could only imagine what having a bath would be like in a storm. There was a vile urine-based stench that I came to understand only after I had eaten a meal or two on board.

There were a couple of hours before casting off so I headed around the port. It was a beautiful Sunday afternoon and people were out fishing along the rocks and wharves. With no clear idea of when I would be back on land, I did what any sailor would do to kill time – I went to the pub.

Later, back on the ship, I watched as the Ministry for Primary Industries' observer, Bheema Louwrens, came on board, but only after a long slow embrace on the wharf with his girlfriend. There was no problem with my taking his designated cabin: his home, a yurt somewhere in Golden Bay, suggested he was relaxed about where he bunked down.

A Port Nelson pilot came aboard. Kylybov was unhappy at having to entrust his ship to him, as he was obliged to do. His worry beads clicked away. Nelson is not an easy harbour from which to exit as it requires a couple of difficult turns, but the *Aleksandr Buryachenko* had done it often enough and quickly. Soon the pilot was happily dropped into a Zodiac and we powered away to the north.

I was assigned to eat with the officers, who were served by a couple of female stewards. Through these officers I learnt the full extent of my Ukrainian: pryvit and spasybo – hello and thank you. I learned that most of the crew came and went on six-month cycles, flown back and forth from Sevastopol on a chartered airliner. Some were in a perpetual winter cycle; Kylybov and others had the summer cycle. The food they had to eat was disgusting, bland and tasteless. It was washed down with a kind of raisin drink that explained the powerful urine smell in my cabin.

We pushed out into the South Taranaki Bight, quickly out of sight of land. For the purposes of the media exercise, *Aleksandr Buryachenko* was to do an exploratory fish for jack mackerel. Few people in the West eat this fish. Despatched headless and frozen, it washes up in some of the poorest parts of the world. It is popular in parts of Africa, where the Koreans and the Japanese have well and truly looted local mackerel, and in Eastern Europe, where it is sold with no adornments. These days, jack mackerel is mainly turned into fishmeal to be fed to farmed high-value fish destined for luxury Western markets.

The business of shooting a net is a kind of ballet involving ropes and large equipment. It has always been dangerous. Heavy ropes and nets go over the stern of a fast-moving ship. Make a mistake and you could be in the water. Sitting alongside the trawl deck but one flight up, I was impressed by the skills displayed by the mostly middle-aged crew. These were rugged men. They were skilled and they knew what they were doing.

On the bridge a couple of officers sat around monitors, watched over by an icon of Saint Nicholas. Some argue that Nicholas, the pre-Coca-Cola Santa Claus, is probably a Christianised version of Poseidon, the Greek god of the sea. I had no problem with his appearance on the ship: just as there are no atheists in foxholes, when at sea you soon learn the importance of higher deities.

As soon as a factory deck and its staff are open for business, a cycle quickly develops of getting something into the net and up on to the deck to keep the factory working. The net is then re-shot. Fishing today involves a lot of computing power and sensor information, and I could soon see that the first mackerel catch was somewhat modest. Despite the stench in the cabin, I went to bed. I've sailed a lot around the world, and even with something of a swell and with her nets out and her engines working hard, *Aleksandr Buryachenko* was a remarkably steady ship.

Early next morning we were up by the Māui gas and condensate platforms south of Mount Taranaki, which was vaguely visible through the hazy horizon. Kylybov had quickly come to the view the area was not good fishing this time and had ordered the nets in. We turned south to go through Cook Strait. Dramatic adventures are possible on this often turbulent body of water but on this summer day it was calm and warm. Kylybov was on the bridge wing much of the time: the strait can be busy with ships and he wanted to keep a close watch.

As the sun set we were off to Kaikoura, the old-fashioned lighthouse blinking out its location. I also had my location, and a note of all other ships around, on my iPhone. Perhaps sensing my disgust with the food, several men offered me bottles of vodka. Rather than drink it, I plunged into my emergency supplies of muesli bars.

Next day, off Banks Peninsula, we did some more mackerel trawling. Again the catches were not great. I went into the factory. It was clean but noisy, wet and dangerous. Fish came down from the pound above and on to a conveyor belt for processing. The bycatch – assorted small sharks and one huge electric ray – ended up to one side. Their fate, dead or alive, was to be turned into mush. I was told the skate was dangerous to touch: I might get a lethal shock.

Louwrens, a larger than life figure, reached into the conveyor to haul out bycatch for scientific study. He then took samples of the mackerel. His most significant work would come later in the voyage, when the ship was fishing for arrow squid at the Auckland Islands, 465 kilometres south of Bluff. This $75 million-a-year squid fishery had one disastrous bycatch: New Zealand sea lions.

Public opinion demanded that sea lions, an intolerably cute species, must be saved. To achieve this, a sled-like device that would supposedly stop sea lions getting caught in squid trawls had been created. One of Louwrens' tasks was to ensure the sleds met specifications. He did the measuring one afternoon on the trawl deck, and later joined me on the foredeck. We shared a passion for watching the deep-sea birds seldom seen on land. These amazing creatures flit across the surface of the sea, constantly circling ships. To my untrained eye, the array of wires around the stern of the *Aleksandr Buryachenko* that was designed to prevent bird bycatch seemed to be working.

Louwrens, a South African-born musician, computer programmer, traveller and circus fire-eater adorned with large rubber washers in his earlobes, had done four observer trips; the rest of the time he lived in his yurt. He had taken the observer's job because he 'got sick of my own hypocrisy. … I spent my life talking about doing stuff to protect the environment and look after the world and it was "blah blah blah". Now I feel I am doing something constructive to help protect the fishery.'

His job involved monitoring the procedures around bycatch, which are governed by complicated rules. Most fish accidentally caught have to be measured against a boat's quota and penalties are imposed. Most bycatch cannot be thrown back, even if alive. Louwrens felt for what the fishing regulations called 'schedule six' fish: worth nothing much to anybody, dead or alive, they didn't have to be kept.

When it came to sea lions, he was well aware of the power of publicity. 'The more attention the politicians and the public give you, the more change you can bring about. If I was trying to save some little ragged-tooth guppies, no one would give much of a shit, even if they were critically endangered. Sea lions are fluffy, big and impressive.' He mused on the irony of the ship's preventing a couple of sea lion deaths while at the same time hoovering up squid, the creatures' main food source.

Louwrens believed the main criterion for becoming an observer was whether you could stand being at sea without going mad or making others mad at you. He had worked on Japanese, New Zealand and Ukrainian boats; only on the New Zealand boat had he encountered resistance.

The presence of observers meant, in the minds of most crews, that the fish stock was being protected. 'I was expecting to be given shit all of the time, and I do know some observers who have been, but in my experience it's almost the opposite,' he said. 'International boats have fished elsewhere. They get paid on a commission basis according to how much they catch and they've seen how thoroughly fucked the rest of the fisheries are. They get peanuts. Then they come to New Zealand. There are lots of fish; they get lots of money.'

Like most observers, Louwrens knew little about the pay and conditions of the ships' crews; it was the responsibility of another government department.

Observing involved a lot of paperwork and sampling. To fight boredom when not working, he read, or wrote 'amusing anecdotes for my autobiography'. While environmental consciousness was partly why he did the job, the pay was pretty good. 'I can work a couple of months and then go home and pick my nose for a couple more and not worry about money. It's padding in my back pocket. Sailing is my passion. I don't have a boat at the moment and this needs to be rectified.'

Later, under the Official Information Act, I would ask the ministry what it paid observers. It refused to say, claiming the contracts were private. I pointed out that we knew everybody else's public service salaries, including that of the prime minister. They told me. The amount did not seem extravagant. After my story quoting Louwrens was published, the ministry fired him. Talking to the media was against the rules.

During the day, Sealord and Kylybov decided, via satellite phone, that mackerel was not paying off and the ship should go south. I could have stayed on board for another six weeks but the food decided it for me. Besides, I had been to the Auckland Islands. A couple of years earlier I had joined a frigate, HMNZS *Te Kaha*, in Bluff and sailed south to them and the Campbell Islands. It had been a rough and challenging trip, but I had seen the wild beauty of these subantarctic places.

At Enderby, a northern island in the Auckland group, we had gone ashore. Sometimes knowing news stories of some time back does not help. As the Zodiac came on to the white sand beach and we had to jump waist-deep into the cold water to go ashore, I recalled that a great white shark had nearly mauled a Department of Conservation worker to death in the same waters.

The beach was crowded with sea lions, most of them fairly aggressive, and a big gathering of pups. I was quickly persuaded of the need not to fish the area. Human survival doesn't depend on how much squid is caught, but its loss could be life or death for these endangered mammals.

It was going to be too expensive for the *Aleksandr Buryachenko* to dock somewhere and let me off. A compromise was reached and, like some nineteenth-century gunrunning operation, it was played out under a gibbous moon at around two in the morning off Aramoana at the entrance to Otago Harbour. Kylybov had the worry beads working; the plan was that he would sweep by the entrance and the pilot boat would swing in and pick me up. Nobody was actually going to stop: I would have to climb a long way down a Jacob's ladder. 'You will be safe, Michael,' Kylybov said. It was both a statement and a question.

I was not worried: the *Te Kaha* trip to the Southern Ocean had taught me the skills of scaling down the side of a ship. That said, the Ukrainian ship was twice the height of anything I had scaled on the frigate. Kylybov gave me a bottle of the finest Ukrainian vodka as a farewell gift. The powerful pilot boat came by and I went over the side. Just at the point I was about to look back (even though the navy tells you never to look up or down), a comforting voice from the pilot boat said, 'Got you, mate.' Forty minutes later I called a taxi and was soon in one of Dunedin's finest hotels.

In 2013 the Geneva-based International Organization for Migration released data on the exploitation of Ukrainian 'seafarers and fishers trapped for their labour at sea'. The study, funded by the US State Department, called it little-noticed human trafficking. Rebecca Surtees of Washington's NEXUS Institute had interviewed 46 Ukrainian seafarers, ranging in age from 18 to 71. The men had been trafficked all over the world – including to South Korea, where they had been exploited at sea for seafood processing. The men had been led through a calculated maze into a world of imprisonment, backbreaking labour, sleep deprivation, crippling and untreated illness and, for the least fortunate, death. 'These men, seeking honest work at sea, ended up on slave ships without means of escape or reasonable prospects for rescue.'[1]

Regulations for fishing vessels were less strict than those for commercial ships, the report pointed out. 'Without exception and regardless of vessel or destination, trafficked Ukrainian seafarers worked seven days a week, for 18 to 22 hours each day. Living and working conditions were extremely harsh. In addition, basic necessities like food and water were universally scarce.'

The report noted that commercial fishing had the unique potential to be exploitative and dangerous. 'The very nature of the work largely out of sight, at sea and, thus, inescapable, and moving between various national and international jurisdictions lends itself to a high risk of abuse.'

Labour and human rights abuses were commonplace on illegal and unregulated vessels. Physical and psychological mistreatment suffered by crews included 'abhorrent working and living conditions; the use of hazardous equipment; inadequate training in the use of equipment on board; long shifts with little time to rest; inadequate medical services, equipment or facilities on board vessels, and forced labour.'

The report stressed that the experience of being trafficked at sea had severe long-term effects. It posed a substantial threat to the identity of the men – on a professional level (as skilled and capable seafarers) as well as on a personal level (as educated and successful men, husbands and fathers).

'Where one's sense of self and identity is intimately tied up in one's profession, having that profession undermined by a trafficking experience can be disorienting, challenging, even debilitating. While this may be most evident for men who have left the profession, it is equally an issue for those who have ambivalent feelings about continuing their chosen profession.'

13 Inquiry

I t was a hot, sticky, classically Samoan night, with the addition of a bit of wind and rain as a nearly formed cyclone died to the north. Backed into the roll-on Apia wharf was *Lady Naomi*, pride of Samoa Shipping Corporation's otherwise modest fleet of interisland ferries.

Our first destination that night in February 2006 was Atafu, Tokelau's northern atoll. Five hundred kilometres from Samoa, Atafu is the northernmost piece of New Zealand in the world. The trip was under the supervision of the United Nations. I was along to observe the 1,200 people of Tokelau vote as to whether or not to gain more independence from New Zealand.

It was going to be a rough 30-hour voyage. I swallowed a seasick pill and headed on board. Up on the top deck, as we prepared to push out into the cyclone-disturbed waters, was the man who would probably be the last governor of Tokelau. Neil Walter stood heroically, the strong wind blowing his hair around his face. 'Goodbye,' he announced. 'I will see you when we arrive.'

Walter had made countless trips to Tokelau, a territory consisting of three atolls without a harbour or an airport. On every trip he suffered the full onslaught of seasickness, a relentless gloomy ailment cured only by hard ground. That night I, too, did not have an especially pleasant time, although I got to spend much of it in my bunk, apart from occasional forays to lose my dinner overboard.

Next morning, most of us – Walter excluded – managed to make our way outside to pass a sunny day watching the South Pacific go by. Occasionally, in the distance, it would be clear that a great school of tuna was in motion. The sea would ruffle; flocks of birds would patrol overhead. Now and again some would plunge into the ocean. We were in one of the loneliest parts of the world but life was all around us.

Early the next morning we arrived off Atafu, and Walter, grey but alive, emerged to take on his semi-colonial duties. We wended our way that week from Atafu to Nukunono and finally to Fakaofu, both taking and observing the voting and ultimately learning that Tokelau preferred to stay with New Zealand.

We were to take our leave the next morning. The previous night the fishermen of Fakaofu had put to sea to catch skipjack tuna, in order to get it, chilled for the Apia market, on to the *Lady Naomi*. Judging by the catch I saw from just one boat in a small amount of time, the waters around Tokelau were rich in fish. Tokelauans don't abuse this. They have the custom of inati, sharing. If a person within the group has some good fortune out fishing, the catch is shared. Fish is important, providing ten to 12 percent of the energy needs of each Tokelauan. The real point of this parable, though, is not about fish. It is about Walter.

By the middle of 2011 the Auckland University Business School team's relentless release of data, together with the publicity I was generating throughout Fairfax Media, had started to tell on the government. One of the problems for the prime minister, John Key, was that while his fisheries minister, Phil Heatley, and his labour minister, Kate Wilkinson, had stoically defended the use of foreign charter vessels, the growing amount of evidence we were presenting was showing things were very wrong.

When boats began sinking and bodies were being brought ashore, as had happened with *Oyang 70* off the South Island coast, Heatley and Wilkinson were forced to do something. In July 2011 they formed a committee. This was not a commission of inquiry but a body just below it in rank, a ministerial inquiry. Five months after setting it up, Heatley lost the fisheries portfolio. (In November 2012, Wilkinson would resign because of her inaction over labour conditions at the Pike River coal mine on the West Coast: it had blown up on her watch in November 2010, killing 29 men.)

The inquiry was headed by Paul Swain, a former Labour member of parliament with little background in the fishing industry but plenty of confidence: when he left parliament in 2008 he had said in his farewell speech that his achievements and successes had been 'far too many to detail' and described his maiden speech as 'a brilliant, visionary piece of work'.

As minister of immigration in 2004, Swain had developed a strategy to engage with operators of foreign charter vessels to resolve 'labour-related issues around foreign crew', saying in a briefing paper to the minister of labour: 'Practices fall short of the required New Zealand standard'. However seven years on, the strategy, whatever it was, had clearly failed.

The second member of the inquiry was Sarah McGrath, a chartered accountant and director of the multinational firm KPMG. The third was the chronically seasick Neil Walter.

The Seafood Industry Council, a powerful group dominated by the big fishing companies, had told the government its members wanted to keep foreign charter vessels. The council had worked up something of a campaign against the business school and me. The central theme was we did not understand business and nor were we patriotic. Announcing the inquiry, Wilkinson told reporters the council was concerned about the publicity the issue was producing.

The inquiry held public hearings around the country. I attended the one at Auckland's Hotel Grand Chancellor. One of those featureless transit hotels, the place's sole merit was that it was near Auckland Airport.

There were some of the usual fishing industry cast of characters around, men who looked uncomfortable in suits and unhappy to see a reporter. I was the only local one present. It is perhaps a reflection on the state of the New Zealand media that I seldom found competitive pressure on the fishing story.

There was, however, an American reporter in attendance. E. Benjamin Skinner had written a 2008 book, *A Crime So Monstrous: Face-to-Face with Modern-Day Slavery*, which had been published by Free Press in New York. As a fellow of the Schuster Institute for Investigative Journalism at Boston's Brandeis University, Skinner had the financial resources to travel the world investigating slavery. And he was on my patch.

I found myself feeling the way many a South Pacific journalist must have felt when I rolled into their country to do a story on a local issue. However, I quickly accepted that as most New Zealand media were not overly keen on the story, Skinner could take it to a much larger international audience.

He was to have a big impact on the way events unfolded in New Zealand, but on the day of public hearings he was a bit like me, sitting on a bench listening to set-piece defences of foreign charter vessels from men with vested interests. At the break I had a short chat with Paul Swain, but did not get the feeling he was bringing much to the table. Catching up with Neil Walter, though, was another matter. It was like returning to Samoa; conversation flowed as if we had parted only yesterday. He had noticed my reporting on the issue and was encouraging enough to say it was making a difference. I think he recognised more than Swain that governments did not like nagging public sores.

The cynical nature of the Korean fishing industry in general, and Sajo Oyang in particular, was evident in the submission to the inquiry by Soon Nam Oh, the so-called 'managing director' of Southern Storm Fishing Ltd, the shell company created by Sajo Oyang. As was typical in the industry, Oh did not show up in person to make her points. In fact, she was incapable of doing so: she was a shell director of a shell company, the wife of Hyun Choi, the Sajo Oyang sidekick who had run Southern Storm Fishing and had moved off the paperwork when *Oyang 70* sank. She was just a shadow, speaking for the Korean company, warning that New Zealand had no legal jurisdiction past the 12-mile limit. 'The EEZ is not part of New Zealand, but rather it is an area over which New Zealand is given certain sovereign and jurisdictional rights and obligations,' her submission declared.

'Parties with broader political agenda' were, she claimed, exploiting the *Oyang 70* sinking. Auckland University and I were acting in a 'politically motivated campaign designed to create a moral panic amongst fair-minded New Zealanders'.[1]

Peter Dawson, an admiralty lawyer from Nelson (not to be confused with Pete Dawson, the local agent managing the Oyang vessels), had dealt with foreign charter vessels in Africa and knew the Koreans well. 'Korean-flagged vessels have a uniformly poor reputation amongst Third World African states for poor labour practices and unscrupulous pillaging of Third World fishing nations,' he told the inquiry.[2]

The submissions from the fishing industry were predictable. Sanford came out against any control over the use of foreign charter vessels; any restrictions would 'seriously hamper the future development of New Zealand's fishing industry'. The company's gruff managing director Eric Barratt said interference with Sanford's business would be 'unwarranted and counterproductive'.

Sealord claimed that if New Zealand gave up on foreign vessels the country would lose $196 million a year: 'FCVs generally harvest low-value species that are unprofitable and/or not practicable for New Zealand vessels to harvest.'

The Seafood Industry Council wanted more low-wage labour, not less. New Zealand-flagged boats, it claimed, could not get local crews. New Zealanders did not like being at sea for weeks at a time, working in uncomfortable conditions, and living in an isolated and enforced alcohol-free and drug-free environment. 'It is not seen as an attractive workplace for many people.'

Using cheap Asian crews, the council claimed, was no different from other New Zealand companies taking their manufacturing to low-wage countries. Many New Zealand businesses had exported jobs previously done in New Zealand to countries where wages were considerably less than the New Zealand minimum wage. Fisher & Paykel, Fonterra and Icebreaker were doing it. Air New Zealand used Chinese crew on its China services and they were being paid less than New Zealanders doing the same jobs.

On the issue of foreign crews not being paid, the council said there was no evidence of this. In fact it was the press who were the bad guys. According to Alastair Macfarlane, the council's general manager, it was the media coverage and not the fishing that posed a risk to New Zealand's reputation. Removing foreign charter vessels from New Zealand would, he said, reduce fish landings and damage the economy. It was such an idiotic submission that the council later wrote to the inquiry asking for it to be ignored.

One of my on-the-record sources in my early reporting of the fishing scandals had been Peter Talley, owner and boss of the Nelson fishing company Talley's. A colleague would later joke that I was 'Talley's reporter' – newsrooms can be a bit rugged.

By ingrained trade union instinct – I had been 'chapel father' at both *The Evening Post* and *New Zealand Press Association* for years – I had struggled with accepting Peter Talley as anything other than the voice of old capital. He controlled a great swathe of food processing, both on land and at sea, and at one point, while I was researching and writing about fishing, the meatworks he owned was subject to industrial action over wages and conditions. I stayed focused on what I was doing because I knew Talley's boats had a reputation for being safe and its workers well paid, with New Zealand work contracts.

Talley's submission to the inquiry said that foreign charter vessels acted in a regulatory and compliance vacuum, leading to undesirable exploitative practices and a distorted playing field for New Zealand-crewed vessels. The foreign vessels had a punitive impact, 'robbing New Zealand Inc. of substantial economic wealth and exposing our industry to significant risk'.[3]

The company highlighted the fact that most of the catch was processed in China using slave labour and then resold into the world markets as 'Produce of New Zealand'. 'This product,' Talley's said, 'is sold in direct competition to seafood caught by New Zealanders and processed by New Zealanders. … Instead of the New Zealand brand being associated with sustainable harvesting and responsible, trusted New Zealand processing, it is at risk of being associated with Third World processing standards and deceptive marketing.'

Talley's dismissed the argument that New Zealanders would not go to sea. 'It only reinforces the state that such boats are in. [Nobody] should have to work in those conditions and certainly not in New Zealand.' The company was puzzled as to why Māori leaders argued for foreign charter vessels when many of their people were unemployed. 'Influential Māori should demand a change in direction that would allow a group of Māori fishermen to demise-charter fishing craft … and crew it with their young people.'

It vehemently opposed the argument that the foreign boats got fish that would otherwise be uneconomic to catch. 'If it is uneconomic to harvest a New Zealand resource under New Zealand labour conditions and costs, then it is not a resource. Blood diamonds and Asian textile sweatshops use the same justification…'

Talley's also presented disturbing statistics. Foreign fishing boats accounted for 90 percent of all seabird strikes, 70 percent of all deep-water offences, and 100 percent of ship desertions.

In February 2012, the joint ministerial inquiry presented its report to the government. It recommended that by 2016 all foreign charter vessels operating in New Zealand be required to reflag to New Zealand so they could be forced to comply with local law.[4]

There were, however, strong signs that the government was going to allow exemptions. Certain Māori iwi had lobbied to be allowed to keep using slave boats because, they argued, this was their right under the Treaty of Waitangi and there were no alternatives. In July 2013, the primary production select committee hearing the bill changed the wording to allow iwi with Treaty of Waitangi settlement quota the right to continue using foreign charter vessels. Ngāpuhi leader Sonny Tau claimed credit for the change, which he said would 'only benefit Ngāpuhi, Sealord Fisheries and Sanford'.[5]

14 Melilla

'This New Zealand story is deeply personal for me as my family is half Kiwi and I have a great deal of affection for your country,' Ben Skinner tells me. 'Once I began looking more closely, the story became even more complex and fascinating.'[1]

Skinner's backer, the Schuster Institute for Investigative Journalism, is the first investigative reporting centre to be based at a university. Launched in 2004, it has pioneered a new breed of non-profit journalism. Its goals are to investigate significant social and political problems and human rights issues, and to uncover corporate and government abuses of power.

Skinner initially dug into the fishing story through the reports coming out of the University of Auckland Business School. I was working on the same story but his outlet was somewhat larger: *Bloomberg Businessweek* boasts of a global circulation of 980,000 and is available in 150 countries.

Skinner is a pescetarian – a vegetarian who eats fish. He had been surprised to discover he could find out little about the direct supply chains of his main protein; he hoped his story would lead to greater openness 'in a shockingly secretive industry'. He focused on two Korean ships, *Melilla 203* and *Melilla 201*, both fishing for Christchurch-based United

Fisheries. In a six-month investigation spanning three continents, he found cases of debt bondage on the *Melilla 203* and at least nine other ships that had operated in New Zealand waters.

One of his informants was 'Yusril', a pseudonym to prevent people in New Zealand and Indonesia finding out who he was (although eventually they did).

On March 25, 2011 Yusril had signed up with the East Jakarta offices of fish crewing agency PT Indah Megah Sari and been offered a job on *Melilla 203*. Skinner outlined the sign-up process as Yusril had described to him. "'Hurry up,' said the agent, holding a pen over a thick stack of contracts in the windowless conference room with water-stained walls. Waving at a pile of green Indonesian passports of other prospective fisherman, he added, "You really can't waste time reading this. There are a lot of others waiting and the plane leaves tomorrow'."[2]

Yusril needed the monthly salary of US$260. His wife was eight months pregnant, and he had put his name on a waiting list for this opportunity nine months earlier. He had paid the agent a US$225 fee, money he had borrowed from his brother-in-law. Other fishermen getting jobs through the agent had borrowed from loan sharks to cover the fee. A few had sold their possessions, including livestock and land.

Skinner reported that the contracts Yusril signed were in English, a language he didn't understand. In the first contract – the 'real' one – the agent had set out the terms. As well as his commission, the agent would retain 30 percent of Yusril's salary until the job ended. Yusril would be paid nothing for the first three months, and if he didn't complete the job to the fishing company's satisfaction – the requirements were left vague – he would be sent home and charged over US$1,000 for the airfare. He had to work whatever hours the boat operators demanded.

The last line of the contract, in bold type, warned that Yusril's family would owe nearly $3,500 if he were to run away from the ship. Yusril had already submitted the title to his land as bond collateral, and had provided the agent with the names and addresses of his family members. As Skinner noted, 'He was locked in.'

Yusril was flown to Dunedin and put aboard *Melilla 203*, a 59-metre-long trawler. It was rusty and dirty, the crew's quarters were damp, and there was no way to keep clean. Nevertheless, it was luxury compared with the boat he had worked on two years earlier, *Dong Won 519*, chartered by Sanford.

'On that boat,' Skinner reported, 'the Korean officers had hit Yusril in the face with fish, and the boatswain had repeatedly kicked him in the back for using gloves when he was sewing the trawl nets in cold weather. Most unnervingly, the second officer would crawl into the bunk of Yusril's friend at night and attempt to rape him.'

As *Melilla 203* sailed into the Southern Ocean conditions worsened. The Korean officers 'became coarser with their language, and would taunt the Muslim crew with bacon. The boatswain would grab crew members' genitals as they worked or slept. When the captain of the ship drank, he molested some of the crew, kicking those who resisted. As the trawl nets hauled in the catch … the officers shouted orders from the bridge. They often compelled the Indonesians to work without proper safety equipment for up to 30 hours straight, swearing at them if they so much as requested coffee or a bathroom break. Even when not hauling catches, 16-hour workdays were standard.'[3]

The long hours meant injuries were common. Ruslan, a crewmate of Yusril, broke two bones in his left hand. It was three weeks before he was taken to hospital, and in the meantime he was made to continue working. After being briefly treated he was abandoned. 'I was a slave, but then I became useless to the Koreans, so they sent me home with nothing,' he told Skinner. He would eventually be paid $335 for three months' work and blacklisted by the agency.

Skinner commented that while *Melilla* ships had been fined or seized for fishing offences, crimes against humanity had taken a back seat. Scott Gallacher, a spokesperson for the Ministry for Primary Industries, which now included Fisheries, had confirmed that observers were 'not formally tasked' with assisting abused crew, although they could report abuses to the Department of Labour at their discretion. The message did not seem to have got through. On board *Melilla 203* Yusril had whispered to an observer, asking for help. He was told, 'Not my job.'

After eight months, Yusril and two dozen other crew members protested to the captain about their treatment and pay. A Labour Department investigator who visited the ship when it was docked in Lyttelton had given Yusril a government fact sheet. It stated that, under New Zealand law, crew members were entitled to minimum standards of treatment and should be paid at least $12 an hour. Instead, after deductions, agency fees, and a manipulated exchange rate differential had been subtracted, they were averaging around one dollar an hour.

The unhappy captain threatened to send the men home to face retribution from the recruiting agency. All but four of the Indonesian crew then walked off the boat and found refuge in the Lyttelton Union Church.

Skinner was curious about the company that was allowing this abuse to happen on ships harvesting its catch. The headquarters of United Fisheries in Christchurch, he reported, featured 'gleaming Doric columns topped with friezes of chariot races … designed to resemble the temples to Aphrodite in Cyprus, the homeland of United founder Kypros Kotzikas. The patriarch started in New Zealand with a small fish and chip restaurant. Some 40 years later his son Andre, 41, runs a company that had some $66 million in revenues last year.'[4]

Although three *Melilla* crew members had fled, citing abuse, just nine days earlier, Kotzikas told Skinner he had heard of no complaints from crew on board the chartered *Melilla* ships, and had personally boarded the vessels to ensure the conditions were 'of very high standard'. 'I don't think that claims of slavery or mistreatment can be attached to foreign charter vessels that are operating here in New Zealand,' the United boss told Skinner. 'Not for responsible operators.'

While New Zealand's labour laws were 'a thousand pages of, you know, beautiful stuff,' Kotzikas believed they did not necessarily apply beyond New Zealand's 12-mile territorial radius.

He told Skinner that United Fisheries sold ling, a species of fish taken by the *Melilla* boats, to Costco Wholesale, the United States' largest wholesaler and one of its biggest retailers; the managing director of Quality Ocean, the Christchurch-based company that exports the fish, would later deny the Costco fish had come from these boats.

Ling can be taken both by trawling and longline, but those caught by line can fetch up to three times the price because the quality is considered superior and the fishing method more environmentally sustainable, something of importance to a growing number of consumers in the West. Because of this there is a temptation for companies to catch ling by trawling and then claim it has been caught by longline. It is a complex operation: catches in one area are declared as trawled and then mixed with longline-caught fish.

United Fisheries also exported squid caught by the *Melilla* boats. Skinner found that outlets for this squid included P. F. Chang's China Bistro, an Arizona-based chain with over 200 restaurants worldwide, and Honolulu-based importer P&E Foods, which had purchased at least 22,000 kilos of squid from United since November 2010.

Skinner's identification of companies buying *Melilla* fish was important, as the United States was moving towards consumer laws in which selling products that had any hint of people-trafficking involved would be regarded as a criminal offence.

'Yusril's story,' Skinner wrote, 'and that of nearly two dozen other survivors of abuse, reveals how the $85 billion global fishing industry profits from the labour of people forced to work for little or no pay, often under the threat of violence. Though many seafood companies and retailers in the US claim not to do business with suppliers who exploit their workers, the truth is more opaque. Beyond the reach of international regulators, human rights violations are committed on a daily basis on the high seas, in the name of satisfying the world's appetite for seafood. This is the story of how that ill-gotten catch may wind up on your plate.'

To complete the circle, Skinner had gone to visit Yusril in his village in Central Java. The former fisherman was back in his in-laws' modest home, out of work, and brainstorming ways to scratch out a living by returning to his father's trade, farming. The recruiting agency had not only blacklisted him but was refusing to return his birth certificate, basic safety training credentials and family papers. It was also withholding his pay, which totalled around US$1,100. Yusril had earned an average of 50 cents an hour working on *Melilla 203*.

A lawyer for the crewing agency did not respond to repeated emails from Skinner. When the reporter showed up at its offices, a security guard escorted him out.

Skinner reported that two of the 24 men who had walked off the *Melilla 203* had returned to work on the ship, rather than face deportation. 'The ship's representatives flew the remaining 22 resisters back to Indonesia. When they returned to Central Java, they say [the agency] coerced them into signing documents waiving their claims to redress for human rights violations, in exchange for their originally stipulated payments of between $500 and $1,000.'

Yusril was one of two who held out. When Skinner last spoke to him, he asked him why he had refused to sign the document. 'Dignity,' Yusril said, pointing to his heart.

15 Sanford

Like anyone who has spent a lot of time on the islands of the South Pacific, I have long been fascinated by drift voyages. These uncharted, accidental journeys are said by some to explain the way in which Polynesians ended up scattered across the Pacific, but I have never believed that. Polynesian voyagers knew what they were doing and where they were going.

I've had to report on drift voyages now and again, and Tokelau – an almost mystical paradise – is for me a kind of drift voyage talisman.

The island group's most ill-famed drift voyage did not have a good outcome. In 1955 MV *Joyita* sailed out of Samoa's Apia harbour with 25 souls aboard, bound for Tokelau. It never made it; in time its semi-submerged empty hull drifted to Fiji. Nothing is known, even to this day, of the fate of the passengers and crew, but the story forever opened my eyes to drift voyages.

In October 2010, three teenage boys on Atafu, Tokelau's northernmost atoll, got drunk, felt romantic about a girl on a neighbouring atoll, stole a boat and motored off to see her. The engine failed and they ended up drifting towards Fiji. A search by the Royal New Zealand Air Force failed to find the boys – 15 year olds Samuelu Peleha and Filo Filo, and 14-year-old Etueni Nasau. They were given up for dead. A memorial service was held.

Fifty days later Sanford's 1957-gross-tonne tuna purse seiner *San Nikunau* was taking a short cut back from Kiribati to Tauranga on New Zealand's east coast. The 19-year-old Italian-built ship was under the wheel of the first mate, Tai Fredricsen. Normally it would have been heading to Pago Pago in American Samoa, but instead it was in waters between the French islands of Wallis and Futuna and Fiji. 'We generally don't take this route,' Fredricsen told me on a satellite phone. 'We were following the fastest line to New Zealand.'

When he saw a little speedboat on the bow he figured something was amiss. 'We had enough smarts to know there were people in it and those people were not supposed to be there.'

The boys started waving. 'I pulled the vessel up as close as I could and asked if they needed any help. They said very much so. They were ecstatic to see us. They were very skinny, but physically in good health considering what they have been through.'

That I was speaking to Fredricsen was amazing, not in a technological sense but because the Inmarsat number for *San Nikunau* had been given me by Sanford's managing director, Eric Barratt. Hard-bitten and a tough supporter of the world of foreign charter vessels, Barratt had been taking a dim view of my reporting on the fishing boats.

The boys' story gained major international attention, mainly to do with the femme fatale angle. A Canadian movie company wanted to turn it into a film but had only a slight knowledge of how to get to Tokelau. I pointed out that it was not possible to fly to Tokelau, and that the occasional boat trips to the island were time-consuming and infrequent. Their enthusiasm withered.

The next time *San Nikunau* was in the news it was for the wrong reasons. The ship ran into strife in American Samoa, a colonial outpost of 56,000 people, which provides a disproportionate number of recruits for the US military, players in the US National Football League, and recipients of federal welfare. The 197-square-kilometre territory became American in 1900 as a result of a scramble between Britain and Germany over Africa and the Pacific. Germany got the much bigger Western Samoa while Washington took the east, centred on Pago Pago harbour.

After the First World War, New Zealand took over Western Samoa, and in 1962 the country became independent, later dropping the 'Western' from its name. Meanwhile, American Samoa remained a sleepy backwater, strictly controlled by the US Navy. Pago Pago did not become the major strategic port that had been envisaged: shortage of land around the harbour prevented major development and its sheltered harbour was, and is, difficult to transit.

When I lived in Samoa in the 1970s, the biggest employers in the region were two US tuna canneries in Pago Pago harbour. These outfits attracted dozens of fishing boats, from big American purse seiners complete with spotting helicopters through to down-at-heel tubs out of Asia.

Every so often an old Asian boat would run aground on the reefs of Tokelau to the north. It was a good place to run aground – there were only 1,200 inhabitants and with no harbour or airport any investigation into the incident was unlikely. The nameless crews would get ashore and be treated well by Tokelauans for a couple of weeks until a ship arrived to pick them up. A New Zealand government official told me it was an insurance scam: it was easier to run the ships aground for a small insurance payment than sail them all the way back to Asia for minimal scrap value. Tokelau's reefs would have to wear the damage.

American Samoa's two large tuna canneries were owned by StarKist of Pittsburgh and Van Camp Seafood Company of San Diego. Van Camp became Chicken of the Sea, and in 1997 it was bought by an investment group, Tri-Union Seafoods. In later corporate manoeuvres, Thai Union Frozen Products of Bangkok, the second largest tuna canner in the world, took total ownership of Chicken of the Sea.

Eighty percent of the workers in Pago Pago's canneries are foreign, many from neighbouring Samoa, where wages are lower. American Samoans like to wave Old Glory, and they like to get away with stuff – such as not paying the US minimum wage to their guest workers. In 2007 the US Congress passed a minimum wage bill that included American Samoa. The hourly rate for workers in the tuna canneries rose from around $3.60 to $4.76, and was to rise by a further 50 cents a year until it matched the rate pertaining in the rest of the US, then $7.25. Since tuna canneries rely on sweatshop labour, the increased wages made the canneries uneconomic. Chicken of the Sea closed down in Pago Pago.

StarKist, however, stayed on. Known for its 'Charlie the Tuna' brand, StarKist traces its origins to a Slav migrant. In 1910 Martin Bogdanovich was fishing off California; by 1918 he had won a military contract to supply canned tuna to troops in Europe. During the Second World War over half of StarKist's output went to American soldiers.

In 1963, the year StarKist opened its cannery in Pago Pago, the company was acquired by HJ Heinz Company. In 2002 Del Monte purchased StarKist, which was by then based in Pittsburgh. Six years later, Del Monte sold StarKist to Dong Won Fisheries. Today this Korean company controls three-quarters of the canned tuna market in its home country. It has a fleet of 36 boats going after tuna around the world. One of Dong Won's assets is a half-share in a processing plant in Weihai, China. The other half of the plant is owned by New Zealand's Sanford.

If investigation of global fishing shows anything, it is that the labels used by fishing companies are largely a fiction. StarKist's owners are plainly Korean, yet in 2013 the company was defending the use of 'Made in America' on its tuna cans.

'What's up with the new StarKist Made in America label?' tweeted Jonathan Gonzalez, a US seafood blogger. Back came a response from American Tuna, a seafood company that prides itself on fishing for albacore tuna using only pole and line: 'Foreign flag vessels, using who knows what gear type, no traceability, canned in Samoa. Hilarious.'[1]

New Zealand-flagged boats operating out of American Samoa with foreign crews are able to catch the so-called 'Made in America' fish. Sanford operated *San Nikunau* out of Pago Pago, offloading its tuna catches to a local cannery. For much of the time between 2008 and 2011 a man called Rolando Ong Vano was the ship's chief engineer. His wages were based on fish offloaded. Another man, James Pogue, also served spells as chief engineer during 2007 to 2009.

Between March 2007 and July 2011 the tuna that *San Nikunau* unloaded at Pago Pago were worth a total of US$24,862,954.89. Its best effort was in July 2008 with a cargo worth US$2,656,741.92. We know these precise figures thanks to the US Justice Department, which presented them to a grand jury in Washington, D.C., alleging that *San Nikunau* was in violation of international and United States law for that entire time. This led to an indictment against Sanford, Ong Vano and Pogue; Ong Vano turned state's evidence and testified against Sanford.

The charges alleged that the ship had infringed the international MARPOL convention, which came into effect in 1983. This convention is designed to prevent discharges of oil and other harmful substances by ships at sea.[2]

On fishing boats, bilge waste accumulates in wells at the bottom point of the ship. The bilge can contain all kind of liquid material, from rain and sea water to assorted oils and fuels leaking from various places on the ship. The bilges have to be emptied occasionally. Ship operators have two options: they can pump the bilge ashore when they are in port; or they can pump it overboard at sea, provided the ship has a functioning oil water separator.

As well as strict rules on how much oil can be in waste going overboard, MARPOL requires vessels to maintain a book in which all disposals of oil residue and discharge overboard are recorded. There are detailed rules about what is to go into this oil record book, and the book is required to be available for immediate inspection. The grand jury indictment alleged that between 2006 and 2011 Sanford had failed to comply. The indictment cited a series of overt acts and listed the days on which *San Nikunau* had dumped oily waste overboard without making a record.

In a formal criminal complaint filed with the US District Court in Washington, D.C., Special Agent Shawn C. Warner of the US Coast Guard Investigative Service said that in July 2011 the service had inspected *San Nikunau* and discovered a system for moving oily waste from inside the ship into the harbour. '[Untreated] water was being discharged directly over the side...'[3]

Chief engineer Ong Vano admitted there had been discharges of oily waste without use of the separator. He believed that around 250 litres of oil a week leaked from the ship's main engine into the bilge.

The case was thorough and detailed. The *San Nikunau*'s first mate Tai Fredricsen, who had been my source on the Tokelau story, denied he had told crew members to lie, as federal officials had claimed, and refused to travel to Washington to testify. The US District Court judge, Beryl Howell, could not make him attend: the writ did not extend to New Zealand. Sanford would later tell the court it had offered Fredricsen and two Filipino crew all-expenses-paid trips to Washington to give evidence.

As is usual in cases where millions of dollars are at stake, there was much pre-trial manoeuvring. Sanford had hired expensive lawyers – New York admiralty and maritime environmental criminal law specialists Chalos O'Connor and Washington D.C.-based Blank Rome – but sometimes their talents were reduced to silliness. Part of the argument centred on the term 'machinery space' in the context of MARPOL. Sanford told Judge Howell this was defined in *Webster's Dictionary* as 'a room (as on a ship) in which the engine is located'. The judge pointed out that *Webster's* had no such definition; Sanford looked like South Seas hicks who could not even read a dictionary.

The case produced a mountain of pre-trial paperwork and could easily fill a legal textbook. In the end it went to trial by jury in Washington, D.C., with seven charges against Sanford Ltd and former chief engineer James Pogue. The US attorney general and assistant attorney general said they were dealing with 'a conspiracy where the crew of the vessel routinely discharged oily bilge waste from the vessel directly into the sea during its fishing voyages since at least 2007'.

Sanford faced a maximum fine of US$3.5 million and forfeiture of the $24.8-million catch taken during the time it was breaking the law. It would, it told the New Zealand Stock Exchange, vigorously defend itself;

it took its responsibilities seriously 'and would never permit discharges of pollutants into the ocean and nor would we obstruct any reasonable investigation into any allegations made against the company'.[4]

The evidence would show otherwise. Sanford was found guilty on six of seven counts, and fined US$1.9 million for dumping oil waste at sea and then attempting to cover up its actions. It was also ordered to pay $500,000 for the benefit of marine sanctuaries in American Samoa, and put on a three-year probation period, during which it could not work inside United States economic zones or operate out of Pago Pago, despite asking the court to recognise that it had 'been a corporate leader in New Zealand in the development and implementation of sustainability practices'.[5]

US assistant attorney general Ignacia S. Moreno didn't mince words. The sentence made it clear that 'companies like Sanford, who deliberately break the law by discharging oil waste into the ocean over a period of years and lie to the US Coast Guard about their activities, will be held fully accountable under US laws'.[6]

Cynically, Eric Barratt said the conviction and fine, which the company would not appeal, would have little impact on its shareholders: 'We have provided for what we expect the outcome to be in the case; there should be minimal financial impact on our accounts.'

In a bid to counter negative media coverage, Sanford's spin doctors created web pages giving the company's account of what had happened at Pago Pago. It was hollow stuff.

I received a handwritten letter asking me to investigate Sanford further. 'Crews are bullied and threatened,' the writer claimed. 'We are on *Nikunau* at present and are paid $20 for a ten-hour day. How is that [fair]? We are fed only dirty reject fish and sometimes we have to catch them ourselves.'

Addressing the issue that had been in court, the writer continued: 'The reason we did not use the oil machine to take out the oil from water was because it was broken down for a long time and company said expensive to fix so it goes into sea. We are sorry about this but not our fault.'

Conditions had been bad on the ship, with no functioning toilets. '[We] had to shit in buckets and put it in sea for six weeks. The smell and mess was not right.'

The crew had been warned not to say anything about Sanford or 'we will be floating in the water'. On land, the writer said, the Sanford culture was the same, with managers bullying them. He ended with a plea: 'Please help us and ask some public questions.'

The letter was anonymous. Short of doorstepping all crew on the dockside, there was no way I could verify it.

Sanford is the biggest fishing company on the New Zealand Stock Exchange, although just one family, the Goodfellows, controls most of its shares. This wealthy clan has a close association with the National Party through Peter Goodfellow, who has been active in the party for over 30 years, a board member since 2006, and president since 2009.

Sanford owes its name to a 19th-century settler, Albert Sanford. According to early company annual reports, Sanford arrived in New Zealand on New Year's Day, 1864. He acquired *Foam*, a former pilot cutter – a fast and easily manoeuvred sailing boat – to fish in the Hauraki Gulf. He smoked the snapper he caught with kauri chips.

In 1870 Sanford moved his family to Pakatoa Island, where he built a fish-curing plant. When snapper in the Firth of Thames at the southern end of the gulf proved to be of poor quality, he moved to Rakino Island, where there was better snapper close by. Demand grew and in 1881 Sanford obtained a building in Auckland's Federal Street. He started buying other fishermen's catches and in 1894 opened a fish market on the corner of Albert and Customs Streets. He was soon engaged in a battle with rival skippers over fish prices and the freshness of the fish, a battle he won.

In 1900 Sanford started the first trawler operation in the gulf, using a schooner, the *Minnie Casey*. Because of its engines the boat was accused of unfair competition, and the government eventually prohibited power trawling in the gulf.

In 1904 Sanford became a limited liability company; Albert Sanford gave himself a salary of £300 a year as managing director and chair of the board. Nineteen twenty-four would be a grim year for the company: its new chair, W. R. Twigg, was killed big-game hunting in Africa and Albert Sanford died. However, the company also listed on the stock exchange. It was New Zealand's only major fishing company.

In the 1920s Sanford began to use Danish seining, or bottom fishing, in which a large net encircles fish and is pulled towards a stationary boat. The method targets fish that live on the seafloor, and there was an outcry when the public realised the company could be damaging fishing grounds in the gulf.

In the 1980s, after New Zealand declared its exclusive economic zone, Sanford formed a joint venture with Marissco Pte Ltd of Singapore to catch lower-valued species and squid. The two companies set up Sanmar Fisheries Ltd, and, using Soviet trawlers, fished the deep water for species such as hoki, hake, orange roughy, squid and oreo dory.

From the mid '80s Sanford itself relied on Japanese and Soviet charter vessels to fish its catch. In 2000 it purchased two American-owned vessels, the *San Nikunau* and the *San Nanumea*.

In its 2003 annual report Sanford reviewed its history. 'From its earliest days,' it stated sanctimoniously, 'Sanford always accepted that it had responsibilities to its employees. These included a safe working environment, job security, good working conditions and equal employment opportunities. But the policy of social sustainability goes further and recognises that there is a mutual benefit in encouraging employee development through on-the-job and classroom education, and company-organised social events.'[7]

In 2012, Sanford featured in Ben Skinner's *Bloomberg Businessweek* article. Eric Barratt had told Skinner there were no labour abuse issues on his company's ships but the American journalist had found a different story. Near the Lyttelton docks, he had interviewed crew members of *Dong*

Won and *Pacinui*, foreign charter vessels catching fish for Sanford. 'We feel like we are slaves,' a *Dong Won* deckhand told him and simulated a Korean officer kicking him on the ground.

The crew's contracts, issued by PT Indah Megah Sari and two other Indonesian agencies, were nearly identical to the ones signed by the *Melilla* crews: they included the same pay rates and false contracts. The crew had also been subjected to doctored timesheets, similar work hours, and intimidation. They risked punishment by speaking to Skinner. Less than a week earlier three *Pacinui* crewmen who complained to the company about conditions on board had been sent back to Indonesia to face the agents.

Much of Sanford's exports to the United States go through Mazzetta Company, a $425-million Chicago-based corporation that is the largest American importer of New Zealand fish. Mazzetta sells fish caught on Sanford's chartered vessels *Dong Won* and *Pacinui* to outlets across the country, including, according to Sanford's then managing director Eric Barratt, the $10-billion Texas-based supermarket chain Whole Foods Market.

Other buyers of Sanford fish include Nova Scotia-based High Liner, which sells products containing the types of seafood caught by the indentured fishermen on *Dong Won* and *Pacinui* to restaurants across the United States. High Liner's customers include Arkansas-based Walmart Stores and Safeways, America's second largest grocery store chain.

Ben Skinner's report had included a statement from crewman Yusril that aboard Sanford's *Dong Won 519* he had been hit by Korean officers and had faced attempted rape. An outraged Eric Barratt responded aggressively. A statement issued in his name to the New Zealand Stock Exchange said the company had identified Yusril. It was having crews and observers interviewed.

Nine hours later in Boston, Skinner received a text from Yusril: 'Strangers at my house, leaving with my family, very scared, please help.' The men, from PT Indah Megah Sari, were demanding that Yusril withdraw statements he had made to Skinner. On Skinner's advice, Yusril reported these events to contacts in the United States Embassy in Jakarta and he, his wife and infant son moved to a safe house.

I was able to speak via Skype to a man who was looking after Yusril. 'They told Yusril they knew everybody who is in the crew,' he said. 'IMS hold the crew list on everybody who stands up for their wages.' He claimed the agency hired criminals and bribed government officials. 'They have much benefit from the blood and sweat of the fishermen. They treat a person like a slave. It is very easy for them.'

'I would like to say this,' he added. 'To all New Zealand companies who are in the fishing business, create your own agencies with people who can be trusted to appoint Indonesian people. Do not get an agency but a person who works for your company.'

I asked Barratt whether he was putting heavies on to the people in Indonesia who had spoken to Skinner. All he would say was that Sanford would commission independent investigators, although they had not yet done so. The company 'would not use any crew manning agents whom we would consider not to be independent'.

Skinner was appalled almost to the point of tears that one of his sources had been confronted for talking to him; he considered such tactics abominable. 'I would say to Eric Barratt, if he wants to attack me, have at it; the reporting in that piece is bulletproof. For agents to be visiting Yusril and intimidating him and his 22-year-old wife and one-year-old son is as despicable as anything I have ever heard of in the corporate world.'

Tom Mazzetta, chief executive of Mazzetta Company, was embarrassed that his company had been named in Skinner's article as an importer of New Zealand fish and, by implication, involved in the scandal. In an open letter to Barratt he wrote, 'Leadership requires taking responsibility and, in light of this information, changes must be made for Mazzetta Company to continue its relationship with Sanford.' Walmart and Safeways also launched investigations.

When crew members leave ships without being paid they are often helped by supporters, most of whom are volunteers: little money is available.

Usefully, the volunteers also collect testimony. One account that came into my hands was from a man who had been on *Dong Won 519*. He had been interviewed in Indonesian, and the English translation was basic. 'Whereas on my imagination was how beautiful and peace New Zealand is and how good work in New Zealand is,' the man said. 'But after I arrived on *Dong Won 519* in New Zealand water, I was very shocked because it was not compatible with the story I have heard from my friend who has worked on New Zealand boat.'

The 'deck boss man' had, he said, punched and kicked him. The work system was very tiring – 'sometimes during hoki and squid seasons I often sleep only three hours a day'.

He had asked to go home but the captain had told him he would not get any money if he did. 'Strong or not, since the crew can work they are not allowed went home less than two years and must finish the contract. At the end I was surrender even though I work like being slaved by Korean.'[8]

In late 2013, a press statement from Maritime New Zealand slipped into my inbox. It revealed that nothing much had changed at Sanford following the US convictions. Sanford had now been charged in New Zealand with illegally discharging oil from the *Pacinui*, failing to notify a discharge, and failing to notify a pollution incident. The charges followed an extensive investigation by the agency since January 2013, including examination of the ship in Timaru, gathering of photographic and video evidence, forensic examination of samples, and interviews with a number of Indonesian crew members.

Before these charges had been laid, 14 crew members had jumped ship from the *Pacinui*, claiming they had been intimidated and cheated by their bosses. Sanford and *Pacinui*'s officers and owners were facing a fine of up to NZ$400,000 and two years' jail.

16 The Southern Ocean

I n 2006 Warner Brothers released an animated movie called *Happy Feet*. The film, set in Antarctica and made at a cost of US$100 million, went on to gross $384 million. The plot revolves around Mumble, an emperor penguin. He has a terrible voice, but he can tap dance, something other penguins cannot do. As Mumble practises alone, a flock of skuas come in to eat him. Mumble buys time by asking Boss Skua about the yellow band on his ankle. He replies that aliens once abducted him.

Having survived the encounter, Mumble joins other penguins on an iceberg, but he is shooed off because of his bad singing. After a leopard seal chases him he ends up with a band of Adélie penguins. He is accepted and does his tap dance, only to cause an avalanche that exposes a hidden human excavator.

Seeking answers, Mumble visits a rockhopper penguin – although rockhoppers do not inhabit Antarctica. This penguin, Lovelace, has a plastic six-pack ring entangled around his neck. Mumble asks if it is from aliens, but Lovelace says there are no aliens, only mystic beings.

Mumble returns to his flock of emperor penguins to find them suffering from a shortage of fish. When he tells them aliens are taking the fish, the community leader, Noah, says the shortage is a punishment on them because of Mumble's dancing.

Exiled, Mumble goes off to find out what is happening to the fish. After he ends up in a Marine World in a faraway country, scientists put a tracking device on him and follow him as he heads back home, where they discover the penguins' lack of fish. Their film prompts a worldwide debate and governments realise they are overfishing in the Southern Ocean. In a happy-ever-after finale, the United Nations bans Antarctic fishing.

The film had a real-life sequel. In 2011 a sick and bedraggled emperor penguin was found near death on a beach at New Zealand's Kāpiti Coast. Common sense says the creature should have been quietly euthanised, but the moment somebody christened it 'Happy Feet' no government official was going to risk such an act.

After the Department of Conservation spent around $30,000 nursing Happy Feet back to life, a millionaire economist, Gareth Morgan, helped pay for the bird to be transported on a research boat to the Auckland Islands where, complete with a satellite transmitter, it was ceremoniously returned to the wild. Shortly afterwards the satellite transmitter went dead; possibly one of the great white sharks for which the Auckland Islands are famous found the temptation of a large fattened penguin irresistible.

In the hype and hullaballoo about the real-life Happy Feet, few made the obvious connection – that aliens really are fishing in the food range of Antarctic penguins and forcing them to search further afield for food.

A decade earlier, when I was visiting Antarctica as an Agence France-Presse correspondent, a group of us were driving from Scott Base to Williams Field, where our RNZAF C-130 Hercules awaited. As we travelled along the ice shelf an emperor penguin wandered by. The penguin was clearly lost. We got out and looked at it from a distance: there are rules against getting close to wildlife in Antarctica. Later a scientist told me it was pretty well doomed. It was a long way from the sea – a large iceberg was blocking McMurdo Sound – and heading in the wrong direction. Sensibly, we did not attempt to 'save' it.

Antarctica has always had a romantic appeal for New Zealanders. Perhaps this is because ours is one of the closest countries to the icy continent – or, as Henry Kissinger is supposed to have said, 'a dagger pointed at the heart of Antarctica'. Or perhaps we share something of the English fascination with disaster exemplified in Captain 'Titus' Oates' fateful (and perhaps apocryphal) statement: 'I am just going outside and may be some time.'

In the days of heroic science in the 1960s, we watched National Film Unit black-and-white documentaries in which scientists hauled up ancient monsters of the deep. We wondered why these creatures did not freeze to death. No one contemplated eating them. For one thing they had such an unappetising name: toothfish.

Later it became clear that these were special fish. Their bloodstream ran with an antifreeze protein. It was discovered that a toothfish's heart beats once every six seconds. This was useful information for understanding human hearts.

Around 1996, toothfish came into my purview as a reporter when a New Zealand government minister issued a statement saying the air force had spotted an illegal fishing boat in Antarctica's Ross Sea taking the fish. A market for them had been found in Manhattan and Las Vegas. Toothfish's white flesh is rich in oil and tastes like cod. In a bid to hide its unmarketable name, increase its appeal, and perhaps ease the conscience of people concerned about such things, it had been renamed. It was now 'Antarctic cod' or 'Chilean sea bass'.

Later, in the southern summer of 1998–99, I flew to Scott Base to cover the visit of a group of environment ministers from around the world. Some of us flew up to Terra Nova Bay, where the elegant wooden Italian base stands on a rocky shore free of snow and ice. Lunch was served. The fish was delicious, and tasted suspiciously fresh. It turned out to be toothfish; presumably the Italian scientists occasionally caught one or two for themselves and their guests.

When boats began to go after toothfish, some basic facts that should have been known weren't. Scientists knew nothing about the fish's breeding patterns, and little about its lifespan. Toothfish are now believed to be mesopelagic, meaning they live in the middle zone of an ocean at between

200 metres and 1,000 metres, although they have been caught as deep as 2,000 metres. They are big fish: adults grow up to two metres long and weigh in at over 135 kilograms.

They are probably the top predator in the Southern Ocean, sharing the honour with killer whales. It used to be thought that nothing predated on killer whales, but there are hints in isotope studies from New Zealand's National Institute of Water and Atmospheric Research that toothfish do. They also consume penguins, leopard seals and large squid.

The world is slowly coming to the realisation that the Ross Sea is the major, and perhaps only significant, habitat for Antarctic toothfish. Time may reveal that hunting and trapping this amazing creature without fully understanding its biology was not a smart thing to do. Science is not even sure where toothfish spawn, or how long they live, other than that it is a long time. By the time all the facts are known, toothfish may have suffered the fate of North Atlantic cod – virtual extermination.

The question is, why catch toothfish at all? The fishery is already struggling. In 2008 a paper – 'Decline of the Antarctic toothfish and its predators in McMurdo Sound and the southern Ross Sea, and recommendations for restoration' – written by three American scientists for CCAMLR, the Commission for the Conservation of Antarctic Marine Living Resources, contained some alarming statistics.

Research had been rigorous. Since 1971 more than 4,500 Antarctic toothfish had been captured, measured, tagged and released by ichthyologists at McMurdo Sound. The largest one caught had weighed 120 kilograms and been 195 centimetres long. The smallest had been five kilograms, the average 25 to 30 kilograms. Based on 17 tagged toothfish that had been recaptured, it had been found that the fish grew about two centimetres and a kilogram a year.

At the beginning of their research the scientists had been able to capture 200 to 500 toothfish annually, but in recent years the number had fallen to zero. And there were other warning signs. Silverfish – the main food of toothfish – had markedly increased in the diet of Adélie penguins.

The report's writers, who included veteran Californian polar scientist David Ainley, concluded that exploiting the fishery had caused 'a trophic cascade as the food web begins to adjust to the disappearance of its most important predator'.

The trouble had started early. Just a couple of years after commercial toothfishing began, the Patagonian toothfish, the warm-water relative of the Antarctic toothfish, was in serious trouble. Faced with its possible extinction, CCAMLR gradually reduced the legal catch from 30,000 tonnes a year in 1997 to between 1,000 and 2,000 tonnes today. In the last decade, legal vessels going after Patagonian toothfish have fallen from 55 to between five and ten.

Quite how a fish stock has been allowed to be hunted to the edge of extinction to feed an élite market is a fair question. The situation has been made even more perilous by the large amount of illegal toothfishing.

The worst of the pirate fishermen come from Spain, in particular from a family-owned Galician shipowning enterprise, Vidal Armadores S.A. This company came to public notice in August 2003 when an Australian Customs and Fisheries patrol vessel, *Southern Supporter*, spotted one of its vessels taking toothfish in Australian territorial waters in the southern Indian Ocean. Ordered to stop, *Viarsa 1* fled. The captain, Ricardo Mario Ribot Cabrera, led a three-week chase through huge seas and icebergs. At one point *Viarsa 1* was lost in sea ice and in danger; *Southern Supporter* directed her to safer waters. As the pursuit moved west, three other boats joined in: a southern African salvage tug *John Ross*, an icebreaker *Agulhas*, and a Falkland Islands-based British fishery patrol boat *Dorada*. On August 28, after 7,200-kilometres, the little fleet surrounded *Viarsa 1* southwest of Cape Town.

The boat was put under arrest and taken back to Fremantle in Australia. It had 97 tonnes of toothfish on board. Australian authorities, having won the world's longest sea chase, then botched the case in court. A jury found the evidence of fishing violations was only 'circumstantial' and that this raised enough doubt to prevent a conviction.

After two years of trials – the criminal one was followed by a civil one when the Australians attempted to retain the vessel – Cabrera and his officers went home free men. 'We are lucky that we come to pass all this time in such a lovely country that is Australia,' the captain told local media. 'It's a beautiful country with such a gentle people. I make a lot of friends here. These two years that sometimes [has been] very, very long. Now this is all over, the jury said we are not guilty, unanimous, so it's all finished now and we hope that next week we will be home.'[1]

Although Spanish-owned, *Viarsa 1* carried the flag of Uruguay, a South American country that has become a base for illegal fishing activity. The boat was eventually scrapped in Mumbai, but not before the tangled web behind its operation had been revealed.

In 2011, the US Center for Public Integrity's International Consortium of Investigative Journalists, which bills itself as one of America's oldest and largest non-partisan non-profit investigative news organisations, published an article on *Viarsa 1*. It had turned out that Vidal Armadores was linked to more than 40 alleged cases of illegal fishing. ICIJ journalists had confronted Manuel Antonio Vidal Pego, the company's co-owner, who denied being a pirate fisherman. 'You can see I don't have a hook, a parrot on my shoulder or a wooden leg,' he said, in what he described as the company's first on-the-record interview. 'We want to erase a story that has never been erased because there's always someone trying to revive it. So much damage has been done by the bad press, we've gone from a dynamic company to nothing.'[2]

Vidal Pego claimed his ships only took toothfish legally, but ICIJ found that government agencies and international regulators had repeatedly pursued Vidal Armadores. The company and its affiliates had clocked up fines around the world of more than US$5 million. At the same time, the Spanish government and the European Union had granted Vidal Armadores at least €8.2 million in subsidies.

In 2005, another pirate toothfishing ship the *Ross*, one of the world's worst, was spotted in Australia's subantarctic marine territory. It was flying the flag of Togo. A West African country of six million people, Togo has just 56 kilometres of coastline and is not a member of CCAMLR. Australia's fisheries minister Ian Macdonald railed, 'They are fishing with impunity and there is nothing anyone can do about it. That's the bit that absolutely angers and frustrates me. We need to change international law so flag states that don't control their vessels should be put out of the market.'[3]

Eighteen months earlier the same boat, then named *Alos*, had been caught – by legal fishermen – illegally fishing at Heard Island, a remote Australian territory in the Southern Ocean.

Following a boat's changes in ownership can be fascinating detective work, but for law enforcement officials around the world, required to be clear about what they're doing, it is a nightmare. The history of this fishing boat, the *Ross*, is an example of how companies operate in the commonwealth of wild fish. Built in 1975, from 1984 to 1998 it was called *Combaroya Tercero III* and owned in Namibia by Paresis Trawling, a subsidiary of Grupo Oya, the company that built the ship which eventually became *Oyang 75*.

In 1998 it was sold to Cormorant Ltd, reflagged to St Vincent and the Grenadines, and renamed *Cap Georges*. It was then chartered to a licensed French operator, a move presumably designed to get it into the French exclusive economic zone around Kerguelen in the southern Indian Ocean. It went through various changes in ownership, although Grupo Oya Pérez was always somewhere in the background. It had 'owners' in Spain, Namibia and Ghana. Its company registration was, at times, in the Seychelles and then in Togo. All through this time, it was plundering toothfish.

New Zealand claims the Ross Sea as its own, although as a signatory to the Antarctic Treaty it has suspended the claim. The presence of illegal fishing boats shows how weak the claim was anyway. New Zealand had no real way of enforcing the law. The best it could do was get legal fishing underway in the hope this would drive away the pirates.

In 1998, to begin legal fishing in the Ross Sea, a new company, SS Fishing Limited, was formed. Sanford owned 50 percent and New Zealand

Longline 50 percent. New Zealand Longline was, in turn, owned half by Talley's and half by Sealord. The powerhouses of New Zealand fishing were targeting the Ross Sea for the first time.

Two vessels, Sanford's *San Aotea II* and Sealord's *Janas*, were despatched to the area, which can be fished only between January and March, and even then conditions are icy. Sanford's annual reports showed that from around the year 2000 the fishing was hard, and not as productive as they had hoped. Toothfish was virtually unknown to the public and had won public attention only when the environment minister, Simon Upton, had issued press statements about the possible threat to the species from pirate fishing. No one knew how big the piracy problem was, Upton said, partly because 'our understanding of overall fish stocks and their dynamics is by no means complete'.

The minister convened a meeting at Scott Base and three journalists, including me, were allowed to cover it. Although toothfishing was top of the agenda, little came of the deliberations.

Growing concern at pirate toothfishing led to a new mission for New Zealand's Ministry of Defence, even though the navy did not have ice-strengthened ships. In 2000 Judith Martin, one of the ministry's platoon of journalists, dished out a feature article describing the work of a P3 Orion patrol over the Southern Ocean. 'Lonely air force and navy patrols are the last line of defence this country has against fish poachers in our southern territorial waters', it began. The head of the Ministry of Foreign Affairs' Antarctic policy unit, Felicity Wong, was said to be 'passionate about protecting and managing the dull grey fish with the unappetising name that is such hot property in glacial southern waters'.[4]

Two years earlier there had been up to 100 vessels, including many with Spanish involvement, targeting toothfish in the waters off South Africa's subantarctic Prince Edward and Marion Islands, and Australia's Heard and MacDonald Islands. Now the fear was that they would head into the Ross Sea.

'The Southern Ocean has had a history of exploitation', Wong was quoted as saying. 'In the 1920s seals and whales were hunted to near extinction, and in the '60s it was rock cod and icefish. It wasn't until the 1970s, when Russian ships started fishing for krill down here, that southern nations such as New Zealand got really serious about preserving and protecting fish stocks.'[5]

The navy's tactical coordinator, Squadron Leader Tim Walshe, added, 'If something goes wrong there are no airfields close by for us to divert to. And if we go down, it's pretty cold out there. We're flying the flag here, and word will get around the fishing community that we're monitoring our patch, not just New Zealand's exclusive economic zone but further afield.'

The truth was that when the air force spotted pirate fishing boats, it could do little more than write a diplomatic note.

By 2004, with illegal fishing hitting the Southern Ocean hard, the industry began to complain about a lack of policing and international urgency. Sanford's Eric Barratt told *The New Zealand Herald* that piracy affected the market, 'particularly for [Patagonian] toothfish, which is high value'.[6] The Washington-based National Environmental Trust found that nearly 80 percent of toothfish sold globally was illegally caught. A newly formed Coalition of Legal Toothfish Operators claimed a network of Spanish-based fishing firms was doing most of the pirate fishing. 'It is hard to escape the conclusion that toothfish poachers have had little trouble finding ways around the measures taken by governments to restrain them in the last five years,' the coalition pronounced. It identified 42 boats that had been involved in toothfish poaching and said there could be more.

Meanwhile, legal commercial fishing tried to keep pace. Twenty-nine boats were, on average, licensed each year to catch Ross Sea toothfish, with half the allowable catch of about 3,500 tonnes taken by New Zealand vessels. The catch earned around $18 million a year in exports for New Zealand, but the going was not easy. In 2002 Greg Johansson, Sanford's deep-water vessels manager, had revealed that the exploratory fishing with three ships, *San Aotea II*, *Janas* and *Sonrisa*, had proven tough. The ships had needed an icebreaker to help them break into the southern part of Ross Sea.

The complex subterfuge involved in toothfishing was exposed in May 2008, when a Namibian-flagged ship, *Paloma V*, docked in Auckland with 98 tonnes of toothfish, 83 tonnes of nurse shark, and 50,000 litres of fish liver oil on board. It was also carrying shark fins from Lord Howe Rise and South Tasman Rise, although it had no right to fish for these.

Paloma V was half-owned by a Uruguayan subsidiary of Vidal Armadores. Its captain, Jose Antonio Paz Sampedro, signed a declaration that he had not been involved in illegal fishing. Ministry of Fisheries officers Phil Kerr and Dominic Hayden inspected the ship and, unusually, cloned copies of its three computers' hard drives. These showed that *Paloma V* had paid for the provisions of pirate boats elsewhere. There were emails detailing the sharing of bait, fuel and crew. There were even photos in which *Paloma V* moved supplies on to a vessel called *Chilbo San 33*, which, when later spotted by the RNZAF in the Ross Sea in February 2011, had a new name, *Xiongnu Baru 33*, and a North Korean flag.

The computers also revealed that *Paloma V* had been using a particular software, and a website called fleetviewonline.com, which the inspectors assumed showed coordination of illegal fishing in the Southern Ocean. Because toothfishing requires considerable logistical backup, the pirate boats need to operate in a fleet and the software makes that easier. Screen captures showed *Paloma V* working with boats called *Belma*, *Eolo*, *Galaecia* and *Hammer* in an area around the South Tasman Rise.

The Ministry of Fisheries decided to deny *Paloma V* unloading rights and report it to CCAMLR as an illegal fishing boat. This led to a High Court hearing in Wellington. Omunkete Fishing (Pty) Ltd, a Namibian company with Spanish owners, said it had purchased *Paloma V* for €7 million – a book entry, with no cash changing hands – and had permission to catch toothfish. The company had not intended to land a catch in New Zealand, but at some point the ship had ended up in the area and sought to do so at Auckland.

The High Court accepted evidence from the Ministry of Fisheries that *Paloma V* and the two companies that owned it had interests in the operations of at least five illegal toothfishing vessels. *Paloma V*, the ministry said, was 'operating as part of a larger fleet of vessels flying different flags engaged in IUU fishing activities'.[7]

Although the ship was detained for eight days and not allowed to unload, it got off with a warning. Since its release, it has been seen fishing in Antarctic waters under the flags of Mongolia, Belize and Cambodia, although its ownership can always be traced back to the Spaniards.

While the case of *Paloma V* shed light on the various ruses used by companies to hide their activities, it also showed how relatively powerless citizens are in stopping the plundering of the seas. It gets worse. Take the vessel *Galaecia*, one of those sharing the fleet-view software with *Paloma V*. *Galaecia* was built in 2002 with a Spanish government subsidy of €1.5 million. Its GPS, which was supposed to allow the Spanish fishing authorities to monitor its operations, was tampered with. Its punishment from Madrid was to be sent into the Southern Ocean to catch toothfish as part of some doubtful scientific programme. It continued to get subsidies until 2008, when, while under investigation by the European Union, it caught fire and sank off the east cost of southern Africa.

Vidal Armadores features frequently in discussions of pirate fishing. At least nine of the company's vessels have been convicted of illegal fishing for shark and toothfish. But, as the lobby group Oceana has ruefully noted, despite Vidal Armadores' notoriety and the formal blacklisting of its vessels by international fisheries management organisations, it has received subsidies of nearly €10 million from the European Union. Even after its convictions, it was provided with subsidies by the autonomous government of Galicia to support construction of a processing factory.

In February 2011, an RNZAF Orion photographed two North Korean-flagged toothfishing boats in the Ross Sea. One, *Xiongnu Baru 33*, was using deep-sea gill nets – banned internationally because 'ghost fishing' by nets that are lost or discarded has detrimental effects on marine life.

The second ship, *Sima Qian Baru 22*, had the usual labyrinthine history. As Vidal Armadores' *Dorita* it had flown the Uruguayan flag before being blacklisted as an illegal fishing vessel. It had known other lives as *Magnus*, flagged to Saint Vincent and the Grenadines, and the previously mentioned *Eolo*, flagged to Equatorial Guinea. CCAMLR, in its charge of illegal fishing, came up with other names: *Thule*, *Red Moon*, *Black Moon*, *Ina Maka*, *Galaxy* and *Corvus*. The list underscored the way pirate ships operate: some names are simply painted on as soon as a patrolling aircraft or ship has passed by.

Vidal Armadores co-owner Antonio Vidal Pego told the International Consortium of Investigative Journalists that his company had sold both *Dorita* and the other ship around 2006. The New Zealand government did not believe him. A diplomatic note was sent. There was no reply. North Korea, meanwhile, does not adhere to international fishing conventions.

Perhaps because the world of flagging and reflagging is so complex to the lay observer, pirate toothfishing in the Southern Ocean seldom attracts much media interest. Now and again, however, the incidents seem not only extreme but an affront to national sovereignty. On December 19, 2010 an Australian Customs vessel, *Ocean Protector*, detected a fishing boat in the Southern Ocean. The boat's home port, Lome, was painted on its stern and it was flying a Togo flag. However, there was no flag of registration on its mast, as required by international maritime law. The call sign on its side identified it as *Zeus*, a banned vessel previously registered in Sierra Leone and Japan and known as *Triton 1* and *Kinsho Maru 18*. A check revealed Togo had removed its registration.

The Australians informed the ship's master they were going to board because of the lack of a mast flag. The crew hauled up a Togo flag. The Australians told them they had been struck off as pirates. They hauled up a Mongolian flag and painted *Lana* along the side of the ship. The master agreed to an inspection but asked for time to prepare. The crew painted over the words *Zeus* and *Lome*, and the boat headed away.

Around the world, consumers have become more sensitive to environmental issues around the catching of fish, not least because conservation of fish stocks is often a strong local political issue. Commercial fishing has taken some notice of this consumer sentiment – accepting, for example, that dolphins should not be caught in tuna nets. But big fishing groups often appear to operate in a way that suggests they hope no one will notice what they are doing.

South Korea, one of the most avaricious fishing nations, is casual about conservation. Delegates at a CCAMLR meeting in December 2011 heard that *Insung 7* – sister ship to *Insung 1*, which had sunk in the Southern Ocean the year before – had caught 339 percent more toothfish than was allowed under its official quota, and had done so intentionally. The Korean government had fined the ship US$1,500. The 35 tonnes of illegal toothfish it had taken were worth half a million dollars.

The New Zealand delegate noted 'this was one of many incidences of non-compliance by the Korean-flagged vessels … and suggested Korea consider reviewing its domestic arrangements to provide for the imposition of more appropriate sanctions on those responsible for vessels flying the Korean flag'.[8]

CCAMLR was not able to list the ship as illegal: its decisions have to be reached by consensus and Korea held out. The Korean edition of the monthly newspaper *Le Monde diplomatique* reported that Insung was notorious for its violation of conservation measures but was protected by the Korean government. The 'true picture of Korea's ugly overseas fishing industry is labour abuse, human rights abuse, violation of international convention, exploitation of marine living resources and underestimation of the value of life,' the newspaper said.[9]

In 2013 the Russian ship *Sparta* briefly returned to view. Two years earlier it had hit an iceberg 1,000 kilometres northeast of Scott Base. With fears of a potentially disastrous oil leak, it had been saved only by the New Zealand air force flying several missions to drop engine parts and pumps, in one of the most expensive rescue operations the country had ever conducted.

After the ship had been repaired in Nelson, it had gone into the Ross Sea over the summer of 2011 to 2012, along with a sister ship, *Ugulan*. Both boats were 25 years old and not ice-strengthened: they had originally been built for use in tropical waters. Sources told me that *Sparta* had returned to Nelson with a small catch, while *Ugulan* had taken a different track, going north into Fiji's exclusive economic zone.

CCAMLR does not outlaw transhipping within the Southern Ocean. However, it requires notice of an intention to switch cargoes between ships. The Russian owners informed the commission, and the transfer of toothfish from *Ugulan* to *Sparta* took place somewhere between New Zealand and Fiji. *Sparta* then bought the fish back to New Zealand, but it too broke down on the way and had to be towed into New Plymouth.

The fisheries ministry certified the catch as legal with the vital Dissostichus catch document, but in reality the operation had been less than transparent. *Ugulan* had probably headed to Fiji in a bid to traffick toothfish through the military regime, only to discover after several days in Suva that Fiji had no authority to issue the particular catch document. The boat had then put out to sea again and met *Sparta* halfway to New Zealand – just inside international waters.

In 2012 Oceana reported that transhipping is one of the most common ways illegal fish enter the mainstream. At least 20 percent of seafood worldwide, with an economic value of between US$10 billion and $23 billion, is caught illegally. Illegal trade in toothfish is five to ten times greater than the official reported catch. 'By mixing in stolen fish, [these fish] take on all of the documentation of the legal fish and are effectively laundered,' Oceana noted.[10]

For scientists based at McMurdo Sound at the southern end of the Ross Sea, commercial fishing for toothfish is seen as little short of a disaster. The scientific catch in the sound, home to the American and New Zealand bases, fell more than 81 percent in little more than a decade. A University of Auckland biologist, Clive Evans, told me that 490 hours of fishing at previously productive places over the 2009–2010 summer 'came up

The 38-year-old Korean trawler *Oyang 70*, which sank in the Southern Ocean in August 2010 with the loss of six lives while hauling in an oversize catch. 'It is a stain on New Zealand's conscience that these ships of shame are allowed to be operated in New Zealand waters,' the Maritime Union said. *Otago Daily Times*

ABOVE Survivors of the *Oyang 70* sinking were rescued by the crew of New Zealand fishing boat *Amaltal Atlantis*. Here, the New Zealanders search for bodies on one of *Oyang 70*'s life rafts. *Talley's Group Ltd*

BELOW A body is hauled on board *Amaltal Atlantis*. *Talley's Group Ltd*

ABOVE Remarkably, Talley's boat arrived on the scene fast enough to rescue 45 men. Here some enjoy a meal on board *Amaltal Atlantis*. *Talley's Group Ltd*

BELOW Indonesian and Filipino crew members from the sunken foreign charter vessel *Oyang 70* queue at Christchurch Airport on their way home. On the boat the men had endured poor, unsafe working conditions, and had to labour up to 18 hours a day with only four hours' sleep. *Fairfax Media New Zealand*

Hyun Gwan Choi, director of the shell company Southern Storm Fishing, made a brief appearance at a news conference after survivors of *Oyang 70* arrived back in Lyttelton. *Fairfax Media New Zealand*

RIGHT The New Zealand Coroners Court inquired into the sinking of *Oyang 70* in April 2012 and coroner McElrea released his findings in March 2013. Here Greg Lyall, captain of *Amaltal Atlantis*, which rescued the survivors, shows the court photos of the operation. He had noticed 'a very distinct separation between the Korean officers and the rest of the crew. [The Koreans] did not mix with or communicate at all with us or the crew whilst on board the vessel.' *Michael Field*

RIGHT TOP: LEFT Oh Gwan Yeol, Sajo Oyang's New Zealand representative, appeared nervous giving testimony at the inquest. His role in the company was unclear and the coroner later described his evidence as 'of limited value'. *Fairfax Media New Zealand*

RIGHT TOP: RIGHT *Oyang 70* agent Pete Dawson of Lyttelton-based Fisheries Consultancy told the inquest that criticism of Sajo Oyang's maintenance of the vessel was 'based on slim evidence and supposition'. He further claimed he had never heard a single complaint raised by crew of any nationality. *Michael Field*

ABOVE The author accepted an invitation to spend time aboard this Ukrainian vessel *Aleksandr Buryachenko*, which fishes for New Zealand quota under charter to Nelson-based Sealord. Sealord is half-owned by Aotearoa Fisheries, whose shareholders include Māori iwi Ngāpuhi (12.63 percent), Ngāti Porou (6.3), Ngāti Kahungunu (5.4), Tainui (4.89) and Ngāi Tahu (4.77). The other half is owned by Japan's Nippon Suisan Kaisha Ltd. *David Leggott/@Nelsonian*

Ministry for Primary Industries' observer on *Aleksandr Buryachenko*, Bheema Louwrens. He was fired by Sealord for speaking frankly to the author. *Michael Field*

ABOVE Fish are hauled aboard *Aleksandr Buryachenko* by the Ukrainian crew. In 2013 the International Organization for Migration released data on Ukrainian seafarers and fishermen. 'Without exception and regardless of vessel or destination, trafficked Ukrainian seafarers worked seven days a week, for 18 to 22 hours each day. Living and working conditions were extremely harsh,' the organisation had found. *Michael Field*

BELOW Fish being processed on board *Aleksandr Buryachenko*. *Michael Field*

ABOVE Korean-flagged fishing boat *Shin Ji*, which was placed under admiralty arrest after a man died on board and the crew complained of atrocious conditions. The ship sat in Auckland's upmarket Viaduct Basin for months after a joint venture between Sajo Oyang and Ngāti Tama Development Trust collapsed. *Keith Ingram/Professional Skipper magazine*

RIGHT TOP Public relations man Glenn Inwood, who was hired by Korea's Sajo Oyang Corporation after its fishing boat *Oyang 70* sank in New Zealand waters in August 2010 with the loss of six lives. *AAP Image*

RIGHT A street in Tegal city on Central Java's north coast. Tegal is home to the majority of Indonesia crewmen on Korean foreign charter fishing boats in New Zealand waters. University of Auckland researchers found the area characterised by low levels of education, high unemployment and poverty. *Glenn Simmons/University of Auckland Business School*

ABOVE Nylon drift gill nets, developed by Japanese companies in the 1970s, wreaked havoc on Pacific fisheries until banned in 1989. Today, such nets are outlawed internationally because 'ghost fishing' by lost or discarded nets damages marine life, but the practice continues. In 2000, divers documented illegal drift gill-netters catching and finning over 4,000 sharks in Mexico's Revillagigedo Island Marine Reserve. *Terry Maas/Seawatch.org*

RIGHT TOP The crew of *Sparta*, a 23-year-old Russian-flagged, California-owned fishing boat, throw lifeboats overboard while the ships flounders in Antarctica's Ross Sea; it had hit an iceberg while hunting for toothfish and had a large hole in the hull. The ship was not ice-strengthened and a multinational effort was required to save it. *United States Air Force*

RIGHT The legal catch of toothfish has been steadily reduced from 30,000 tonnes a year in 1997 to between 1,000 and 2,000 tonnes today after the prized fish was nearly hunted to extinction. However, it remains a target for pirate vessels, especially from Spain; nearly 80 percent of toothfish sold globally has been illegally caught. In restaurants, to disguise its identity, it goes under the names 'Antarctic cod' or 'Chilean sea bass'. *Daniel Beltrá/Greenpeace*

Parrotfish, found around tropical reefs, munch coral to get at the algae inside and then excrete the ground-up remains, forming white sandy beaches. With sea level rise, it is becoming increasingly important for islands and atolls to protect the species. *Shutterstock*

with a single little fish. … This is telling us something'. He questioned the models developed by CCAMLR that formed the basis of allowable catches of toothfish; under them there would be a reduction in original fish numbers of up to 50 percent over 35 years. 'If we haven't got the models right and if this is now significant, we are in trouble,' Evans said.[11]

Like others, he was astonished that commercial toothfishing had been allowed to take place when so little was known of the species' life history and numbers. Toothfish is thought to breed near seamounts in the Ross Sea, and Evans suspected that as fishing removed part of the main stock, toothfish at the periphery – such as at McMurdo Sound – had begun to retreat, with impacts on the local ecosystem.

One of the world's leading experts on penguins, California's David Ainley, was angry at toothfishing taking place so near penguin colonies. 'Every New Zealander I talk to who is not related to the fishery is upset at what is happening in the Ross Sea, given the tradition New Zealand has had,' he told me. 'If the New Zealand public knew about this, there would be major pressure against the continuation of this fishery.'[12] Ainley believes toothfish are the main predator of penguins, as well as consumers of silverfish. With fewer toothfish, there are now more silverfish and penguins. Meanwhile, toothfish-eating killer whales are decreasing in the Ross Sea.

Just how little is known about the region's marine ecology had become evident in 2007, when Sanford's *San Aspiring*, out toothfishing, hauled up a 495-kilogram colossal squid. When the squid was dragged up, it was gamely hanging on to a toothfish.

There is another unintended side effect of toothfishing in Antarctica. The Ross Sea has provided 50 years of comprehensive climate data and is the best place in the world to work out what is happening with global climate change. However, its usefulness for this scientific work has been imperilled by boats taking too many fish. 'It is no longer easy to separate climate effects from fishing effects. Fishing has destroyed the science,' David Ainley told me.[13]

Meanwhile, Peter Bodeker, chief executive of New Zealand's Seafood Industry Council from April 2010 to November 2012, defended New Zealand's role in the toothfishing. 'The fishery makes up a relatively small part of human activity in an area that has seen continuous human habitation for some 50 years,' he said. He argued that while there were 15 'relatively small' fishing vessels in the area for three months each year, the place was also visited by numerous transportation vessels, aircraft, tourism operations, coast guard and naval vessels. 'It is unreasonable to single out New Zealand fishing activity, given that the area is governed by a regional entity that annually allows fishing by vessels from a wide range of countries.'[14]

While I was completing this chapter, a New Zealand filmmaker called Peter Young was pounding the world with his documentary film *The Last Ocean*. Young's dream is to see the Ross Sea turned into a marine protected area, leaving the toothfish unhindered by man and contributing to the survival of the world's last largely untouched ecosystem. Many thousands of people have joined the campaign, which began in 2006.

After initially opposing the idea of a marine sanctuary, the New Zealand government found the United States government was interested. Keen to maintain at least partial access for toothfishing, it pushed for a limited area to be protected. In 2013, meetings of CCAMLR in Brussels, Bremerhaven and Hobart all failed to come up with an agreement, primarily because Russia – a source of pirate boats – opposed it. Meanwhile, the legal plunder of a species only the very rich can afford to eat continues.

So, too, does toothfish piracy. Purple notices issued by Project Scale – a new Interpol unit set up in 2013 to detect, suppress and combat fisheries crimes – over *Snake* aka *Berber*, a vessel blacklisted by CCAMLR that has actively fished for toothfish in the Southern Ocean for approximately ten years, and *Thunder*, another offender, flagged to Nigeria, appear to have borne no fruit.[15] Neither vessel, in whatever guise, has been arrested. According to Interpol, *Snake* has operated under 12 different names in the last ten years, and been flagged to at least eight different countries.

17 The Pacific prize

During much of 2003 I had been in Suva, covering the wash-up of George Speight's failed coup three years earlier. On weekends I'd sometimes wander down to the wharves to indulge my mild case of ship-spotting. All through Fiji's political dramas life had continued as usual on the dozen or more old Asian fishing boats that languished in Suva harbour among the rusted island ferries and the odd cargo ship.

I cannot say I remember Taiwanese-flagged *Lih Fa*, owned by An Ho Chen Fisheries of Kaohsiung, Taiwan. The 32-metre vessel longlined for tuna south of Fiji and into the Tasman Sea. Early on April 17, 2003, *Lih Fa*'s 65-year-old skipper Chen Ching Kung was fishing in six-metre seas about 800 kilometres west of Auckland when things went wrong. He radioed to another Taiwanese boat, *Loung Dar*, one of seven other fishing boats in the area, saying he needed to go on deck to check things. As the other boats listened, Chen came back on the air: his ship was flooding and he needed help. Then the radio communications cut out. The ship's emergency radio beacon was not activated, and no proper distress signal was given. The ship was carrying 18 people – Taiwanese officers and Chinese and Vietnamese crew.

The boats that arrived on the scene found some floats, oil and fishing gear, but the New Zealand rescue coordination centre was not alerted for 15 hours. At that stage an Orion was scrambled, but in eight hours of searching the pilots saw only an orange buoy. Maritime New Zealand believed they had the wrong position and gave up searching. 'We believe the vessel either swamped or foundered in the pretty horrible seas,' rescue officer Ray Parker said.

At the time there was a small band of foreign correspondents based in Auckland and one of them, Australian Broadcasting Corporation's Gillian Bradford, became angry at the way Maritime New Zealand had handled the search for *Lih Fa*. In a piece for Radio Australia, she noted it had taken a week after the captain had radioed for help for authorities to alert the public to the drama taking place in the Tasman Sea. 'The Rescue Centre's explanation for this doesn't offer much comfort,' she told her listeners. 'It had simply decided a Taiwanese fishing vessel missing with 18 crew on board wasn't newsworthy.'

John Lee, the operations manager of Tai Fi Shipping Agencies, handlers of the *Lih Fa* when it was in port, later said it was probable that sea water had flowed into hatches and flooded the boat, causing it to sink. 'It would have happened so fast. The time for the captain to muster his crew was very short. Everyone probably got tangled in the mess of things because they were working at the time.'

Three months later the *Loung Dar* sank near the Kermadec Islands. This time the crew were rescued.

The *New Zealand Herald*'s Eugene Bingham wrote that the two sinkings raised serious questions: 'What is going on in the waters off New Zealand? Just what kind of life is it for the men who sweat and toil to bring you the cans of tuna in your pantry? Are New Zealand taxpayers paying the price for unsafe practices among the high-seas fishing fleets?'[1]

Bingham homed in on the Fiji fleet. 'Licensing in some places is corrupt and it's a pretty vexed issue,' a source had told him. Safety standards were variable: 'Some of them, sailing out of Fiji, are floating shitboxes; others are seaworthy and sophisticated vessels.' The Mission to Seafarers told Bingham that conditions on foreign fishing vessels were 'awful' and the pay lousy.

Boats, of course, sink all the time and crewmen are mostly seen as expendable in the cynical world of commercial fishing. But there is a wider issue: if this is what fishing companies do to their own people, can they be trusted to exploit the world's oceans sustainably?

I have spent a large part of my life in the South Pacific. I have sat through countless international forums as people have debated how to make money out of the ocean and how to protect it. The growing pessimism I have felt about the fishing of the South Seas began during my early years as a correspondent with Agence France-Presse, when I covered the drift net debate.

Drift gill nets, also known as drift nets, are among the simplest and oldest methods of fishing. A buoyant float line at the top of the net and a weighted lead line at the bottom hold the netting vertical in the water column. The fish are caught passively; sometimes the nets are left at the end of the boat operating them; other times they are left drifting and recovered later.

A study by the Food and Agricultural Organization of the United Nations reckoned drift nets had been in continuous use around the world for millennia. Herring had begun to be caught by drift nets in the North Sea by the eleventh and 12th centuries. In the 16th and 17th centuries the Dutch developed large industrial vessels for drift netting in the open sea and processing the catch on board. By 1908 it was estimated that more than half a million tonnes of herring were being taken annually by drift net fleets in the North Sea. In the autumn of 1913 there were over 1,700 drifters operating from just two English ports, each deploying around three kilometres of netting, eleven metres in depth, in the south of the North Sea every night.

Drift netting was seen as efficient, selective and mostly environmentally benign. The nets were targeted at schools of pelagic fish found in dense concentrations at certain times of year, and it was recognised that the nets' meshes allowed smaller fish to swim through.

Drift nets were used mostly at night as it was harder for fish to see them. Then in the 1970s the Japanese developed nylon monofilament nets for deep-sea drift netting, creating the most destructive fishery yet devised. Fish couldn't see the nets day or night. And because the nets weighed little they could be much longer than the older style drift nets; some in the albacore tuna fishery were up to 55 kilometres long.

Japan, Korea and Taiwan quickly adopted the new nets in the north Pacific, targeting Pacific salmon, neon flying squid, and various tuna and tuna-like species. For Japanese and Taiwanese tuna boats, drift netting also promised big catches in the South Pacific.

Nylon does not easily decay, and with large amounts of the netting deployed it was inevitable large chunks of it would break loose and drift off, fishing uncontrolled for an indefinite amount of time, catching everything in its way. Eventually, a media term galvanised public concern on an issue that had been going unnoticed. Photos of 'wall-of-death fishing' showed not only dead fish in the nets but a massive bycatch of turtles, sharks, dolphins and even an occasional whale. Some of the photos became virtual trademarks for the environmental movement.

The issue caught the attention of New Zealand's Labour government, and in particular the prime minister, Geoffrey Palmer. At the time, South Pacific leaders were looking for a post-nuclear-testing issue to unite them. Wall-of-death fishing did that. On Tarawa in July 1989, during the annual South Pacific Forum, the matter was discussed, and led to the adoption of a new international agreement on November 24 the same year.[2]

I remember going to the crucial meeting in downtown Wellington. Palmer, already under stress and fearing his government would fall at the next election, gave the drift net fishing issue his total commitment. He had a reason close to home: the best known drift net fishery in the southern Pacific was an albacore tuna fishery in the Tasman Sea and waters east-southeast of New Zealand between 30 and 45 degrees south in the subtropical convergence zone.

The Japanese had found and explored this subtropical fishery from 1982. At first there were up to 20 Japanese vessels but in the 1988–1989 season, after the United States ordered drift netters out of its exclusive economic zone, 64 Japanese vessels and as many as 130 Taiwanese vessels hit the area. Their nets averaged 40 kilometres in length, and in the Tasman their biggest bycatch was common dolphins.

One study found that, on the assumption that 20 vessels operated in the Tasman Sea in the 1989–1990 season, each setting 40 kilometres of netting per day for three months, there would be a total catch of some 4,600 dolphins, 6,300 billfish, 3,500 sharks, 2,700 sunfish and 13,800 Ray's breams, as well as 900,000 tuna.

In 1989, armed with the evidence that drift netting was dangerous, Palmer came up with the Convention for the Prohibition of Fishing with Long Driftnets in the South Pacific. Known as the Wellington Convention, it banned the use of drift nets over 2.5 kilometres long in the South Pacific, and paved the way for a global moratorium on drift net fishing on the high seas. Palmer, a constitutional lawyer, may regard much of his other work as career high points, but in terms of human history and survival of the natural environment his restriction of drift netting was his best and most lasting legacy.

In 2013, I met the Sealord chief executive, Graham Stuart, at a hotel at Auckland Airport, and after running through the pluses and minuses of the company we talked about wider issues. I said there was a suspicion that fishing companies were philosophically inclined to hunt down the last fish in the ocean. Stuart made the point that if Sealord were to do that it would cost it $350 million a year – the value put on their quota to catch fish. This sounds rational, but there is something else that drives fishing companies to become reckless, as they had been with the drift nets. The plight of the southeast Pacific's jack mackerel, more than anything else, demonstrates this.

Jack mackerel makes up one of the world's largest marine biomasses, and for a long time it swam in rich waters of the South Pacific, unclaimed and uncontrolled by any nation. Eventually, in a bid to control the fishing, the South Pacific Regional Fisheries Management Organisation was created. From 2006, when it was formally proposed to set up this body, fishing companies panicked, and soon the race for jack mackerel was on.

As documented by the International Consortium of Investigative Journalists, two-thirds of the jack mackerel stock was ransacked. One South Korean ship, Insung Corporation's *Kwang Ja Ho*, overfished its quota by 68 percent. South Korea's Ministry for Food, Agriculture, Forestry and Fisheries suspended the vessel's licence for 30 days and issued a 'correctional fine' of one million won – about US$925. The excess 2,219 tonnes of mackerel the ship had caught were worth around $1.7 million.

Even in the face of compelling evidence of imminent disaster, countries refused to act. In Chile, the country closest to the fishery, self-interest kicked in. The ICIJ found that a handful of companies had secured quotas that were much higher than scientists said they should be if the fish stock were to be preserved. Controlled by wealthy families and backed by the government, these companies together owned rights to 87 percent of the Chilean jack mackerel catch.

New Zealand international lawyer Bill Mansfield, the chair of the management organisation, faced a grim job as fleets of boats from not only Chile and other South American countries, but from Russia, China, Korea, the European Union, the Faroe Islands, Vanuatu and the Cook Islands took what mackerel they could get away with.

The European Union proposed a catch limit of 300,000 tonnes but fishing industry interests succeeded in getting this lifted to 438,000 tonnes and the pillage continued. In 2012, a scientific report to the organisation's annual meeting showed jack mackerel stock levels were at between eight and 17 percent of what they would have been had no fishing taken place.

One day early in 2013, the world's second largest fishing boat, 9,500-tonne, 142-metre-long *Margiris*, stopped off at Nelson around midnight. The Lithuanian-flagged ship did not dock. A quiet transaction went on in the dark. The ship then sailed off through Cook Strait and on to Chile and the jack mackerel stocks.

The year before, *Margiris* had achieved the rare notoriety of being one of the few ships actually banned for robbing West African fish stocks.

Seafish Tasmania, which bills itself as 'a quality company supplying quality Australian seafood', had then had the bright idea of bringing the ship to Australia, where it would be reflagged and used to take 18,000 tonnes of jack mackerel in the Tasman Sea and the South Australian Bight.

The plan created uproar. Before the ship could leave for Australia, Greenpeace managed to delay its departure from the Dutch port of IJmuiden for five days. The Tasmanian Conservation Trust pointed out that the fish *Margiris* would target were a vital food source for important species, such as the critically endangered southern bluefin tuna, seabirds, marine mammals and game fish. Martin Haley, vice-president of the Tuna Club of Tasmania, highlighted strong opposition to the 'ocean-going vacuum cleaner's impending arrival'. Radical environmentalists Sea Shepherd said Margiris's impact would be huge. 'If overfishing does not stop, the world's fisheries will completely collapse by 2048. … allowing this super-trawler to operate in Australia's waters would be a further sealing of humanity's fate'.[3]

With so much adverse publicity, bringing *Margiris* to Australia became politically unacceptable. By the middle of 2013 the boat was instead working jack mackerel stocks out of Chile, where estimated stock of the fish had fallen from 30 million tonnes to just three million. Much of the mackerel caught is likely to have gone into fishmeal to feed farmed salmon.

While mackerel is a comparatively low-value fish, at the other end of the spectrum tuna offers huge rewards because of customers interested in high quality and willing to pay for it. Tuna is the great international treasure on offer in the Pacific. The western and central Pacific is the world's biggest tuna fishery and its catch the prize for any nation in the fishing business, from small coastal communities around the region to Spain and Norway.

For Pacific communities, knowledge of tuna has not been particularly extensive. Tuna is an open ocean fish. For most Pacific peoples it has been other fish, caught in lagoons and reefs, that have sustained life.

In the 1960s, Professor François Doumenge of France's University of Montpellier carried out research for what was then the South Pacific Commission and is today called Secretariat of the Pacific Community. He reported that it was not until 1950 that the successes of the first expeditions of Japanese longliners grouped around mother ships showed that equatorial Pacific waters might offer vast possibilities as fishing grounds for fish of the tuna family. 'However,' he said, 'it was only after the conclusion of the peace treaty between the United States and Japan in 1952 that tuna fishing fleets were able to fish the waters south of the equator. Hence it is not much more than ten years since the territories of the South Pacific were suddenly confronted with a new form of activity which sometimes took place in full view of their shores.'[4]

At first the ships would spend only three or four months in the region and then head back to Japan with their catches. This led to interest in the possibilities of shore bases: these were established at Pago Pago in American Samoa in 1954 and at Pallicolo in Espiritu Santo, Vanuatu, then part of the New Hebrides, in 1957. Policing and supervision of the Japanese fishing activities were non-existent and impossible. The result was what Doumenge called 'the very understandable tendency among Japanese fishing skippers to believe themselves masters of the sea and independent of the jurisdiction of the island territories'.

Even after the Japanese boats moved into the inshore fisheries, there was a surprising ignorance about these developments among Pacific political leaders, and at village and atoll level. What was known in one place was not known in another. Doumenge insisted that Pacific countries had to act together: 'The South Pacific islands, occupying as they do the world's largest ocean, must endeavour to put to good account the wealth of the seas that surround them.'

In the nearly 50 years since those words were written, ships from around the world have taken Pacific tuna on an industrial scale. For a while American boats, many linked to Mafia-controlled operations seeking to smuggle cocaine and heroin on the boats, dominated.[5] But even the Mafia find it hard to compete with the Chinese: in 2010 China and Taiwan took 53,000 tonnes of the total 75,000-tonne Pacific albacore catch. Most of the tuna is now taken by the 300 vessels registered under the flags of these two countries. But as with toothfishing and illegal fishing operations, reflagging has become the name of the game. There is a growing fleet of vessels reflagging or chartered to the Solomon Islands, Vanuatu, the Marshall Islands, the Federated States of Micronesia, Fiji, Cook Islands, Papua New Guinea and Kiribati.

Technology is affecting much of the fishing. Catching yellowfin and big eye tuna used to be limited by the fact the catch could not be snap-frozen fast enough. Now the industry has a new class of refrigerated 'super containers' and 'magnum containers' that allow products to be kept frozen up to minus 600 degrees. These fit on small container ships and can be sent across the Pacific to smaller ports, changing the economics of albacore tuna and making it more worthwhile to catch.

Small longline vessels can be built for US$1 million, as against the $7.8 million for vessels larger than 40 metres. Vessels less than 24 metres long can have a hold capacity of 130 cubic metres, and can set in excess of 3,000 hooks daily. These vessels can be mass-produced in fibreglass or steel and fitted with high-speed industrial main engines. This has allowed the creation of large efficient fleets that are highly mobile and can relocate across the globe at short notice. As the World Wildlife Fund succinctly put it: 'This new logistical and vessel infrastructure and the related changing fleet dynamics point to there being nowhere to hide for South Pacific albacore.'[6]

Observers on board fishing boats are often the only eyes and ears a country has on its exclusive economic zone. However, observing can be a rotten business, as New Zealand helicopter pilot Brian Grant found. Grant, who worked on vessels in the northern Pacific, told the *Otago Daily Times* many observers were corrupt, taking bribes of up to $100 a day to stay in their cabins. On one Taiwanese vessel, the observer and the captain disagreed over the size of the bribe. 'The observer said, "I'm going to tell on you,"' Grant recalled. The captain left immediately for Taiwan.

Grant found that some observers were good, but most were 'lazy, incompetent or corrupt … a waste of food. The last one I had was on an American boat. I think we had been at sea for three weeks before I even saw him. He was drunk all the time.'

Poor observers mean poor statistics, and so the various bodies monitoring fish stocks often have no real idea of what is going on. 'It's just all,' Grant said, 'made-up stuff.'[7]

On Korean and Taiwanese boats Grant often saw high-grading of tuna or dumping of inferior catch for better catch, and sharks being finned and thrown back. Turtles caught would be eaten, as would the occasional small whale caught in a net. It is relatively easy to avoid catching whales by not putting out nets when they are around. Grant, who flew spotter for boats, said the Americans would never let their nets out when whales were in the area but Taiwanese and Korean ships would take a chance.

Another chilling observation was that while Taiwanese boats were built for a crew of 36, on long tuna patrols they would carry 42 – 'spare crew members for when people got hurt or killed. They expected to lose four a year.'[8]

Tuna roam far and fast across the Pacific. In May 2011 a yellowfin was caught in the Bismarck Sea, off the north coast of Papua New Guinea. It was tagged by scientists of the South Pacific Commission and released. A year later, Johnathan Joul from near Madang in northern Papua New Guinea was working aboard a Filipino fishing boat, *Dolores 828*, when it hauled a tuna aboard. He spotted the tag and sent it in. Because it was the 50,000th tag returned to the commission, Joul was awarded US$500.

The data revealed the fish had swum about 260 kilometres from its release point a year earlier. Where it had been was something of a guess: yellowfin are highly migratory and commonly swim entire oceans in their annual travels.

In 2013, a 100-kilogram bigeye tuna was recaptured 1,000 kilometres east of Fiji nearly 13 years after it had been caught, tagged and released. The yellow plastic tag was found by Samuela Ratini, a crew member aboard a Taiwanese vessel, *San Sai FA 12*, and recorded by the ship's observer, Sitakio Semisi from Tonga. Where the bigeye had travelled was also a mystery: it was in pretty much the same area where it had been tagged.

The tuna catch has been rising because of the increasing use and sophistication of purse seine fishing, which targets primarily skipjack tuna. In 1960, 466,110 tonnes of tuna were taken out of a wide Pacific area. In 2010, the catch of the four main Pacific tuna species reached 2.4 million tonnes, the second highest annual catch on record: only 2009 had been higher at 2.46 million tonnes. The 2010 catch was valued at between US$4 and $5 billion. It was 83 percent of the total Pacific Ocean fishing catch and 60 percent of the global tuna catch. Sixty percent of the nearly three-million-tonne take was skipjack.

The tuna business is driven by a great variety of food tastes and desires. No one in Japan has ever starved from not having Pacific bluefin tuna to delicately slice into sashimi and sushi, but restaurateurs are willing to pay ridiculous prices for it. Bluefin's flesh is the darkest. Because of its high fat content, many say it cannot be cooked as it produces a strong fish taste – having fish taste like fish is strangely undesirable.

Highly migratory bigeye tuna, which is reddish in colour and has high fat content, is favoured for sashimi and fetches the highest prices in the vast Tokyo fish market. Although it too is often canned, it is also sold fresh and frozen.

Skipjack is sold primarily as 'canned light tuna' and is said to have the most pronounced flavour of the tropical tunas. Yellowfin has a mild, meaty flavour and is firm and moist. Yellowfin is sold fresh, frozen or canned as a light-meat tuna, and is served raw as sashimi and sushi.

Albacore is a softer, milder form of tuna, not really suited to sashimi, but with the highest levels of omega-3 fatty acids, believed by some to prevent heart attacks, fend off arthritis and even boost brain power, although none of this has been conclusively proved. Albacore mostly ends up canned as 'white meat'.

A feeder at the top end of the marine food chain, tuna concentrates mercury in its flesh, although how so much mercury gets into the ocean is still largely unknown. Smaller, younger albacore caught with troll and pole-and-line gear usually have lower mercury levels than larger longline-caught albacore, but how a pregnant mother, or anybody else wishing to avoid mercury, can obtain that information from a can is anybody's guess.

18 The China syndrome

With nearly 6,000 men and women aboard, the 101,300-tonne nuclear-powered aircraft carrier the USS *Carl Vinson* did not appear an ideal fisheries patrol craft. Toss in the rest of its strike group, including cruiser USS *Bunker Hill* and destroyer USS *Halsey*, and the whole operation seemed a noisy and expensive way to check fishing-boat logbooks and net mesh sizes. But returning from the Arabian Gulf in May 2012 these ships did fisheries work, and with good reason.

Carl Vinson sailed to Western Australia and then, heading for Hawai'i, operated through the north Tasman Sea and across the South Pacific. Its orders were to 'patrol and secure protected fishing areas'. The semi-official *Navy Times* saw this as a sign the US Navy was expanding its Pacific mission. 'A key area,' the paper noted, 'is what's known as the "Tuna Belt", which runs along the equator and supplies 57 percent of the world's tuna.'[1]

US Coast Guard Commander Mark Morin was quoted as saying that each year about US$1.7 billion was lost in revenues to licensing governments and legal fishers due to illegal fishing in the South Pacific; depleted stocks in other areas of the world were forcing boats into the region. Captain John Steinberger, commodore and commander of the San

Diego-based Destroyer Squadron One, said having an aircraft carrier and its two escort ships headed to Hawai'i on separate tracks increased the breadth of surveillance and patrols over an area stretching 480 kilometres.

Neither man stated what was found, but it was clear from the way they spoke that they had plenty to do. 'I suspect as this mission becomes more widely known, we'll be able to do some more creative things, like plan our tracks through some areas where intelligence would indicate there'd be more illegal fishing vessels,' Steinberger said.[2]

In 2003 Michael Lodge, an English lawyer, helped set up the Western and Central Pacific Fisheries Commission, an international fisheries body that seeks to ensure, through effective management, the long-term conservation and sustainable use of highly migratory fish stocks – that is, tuna, billfish and marlin; he then joined the Marine Stewardship Council. Lodge tried at one point to quantify illegal fishing in the Pacific, which he 'expected to be present in most Pacific Island country exclusive economic zones, where enforcement is lacking'.[3]

Under-reporting and non-reporting were also significant, he said. 'Efforts to prevent such practices as transhipping at sea by requiring vessels to enter port to tranship have reduced the extent of non-reporting, but it is still present.' Tuna pirates meet shadowy cargo vessels on the high seas and transfer their catches to them; the fish are then taken to Japan where, depending on the state of quotas and placement of legal ships, the cargo ships declare the fish have come from the south Atlantic or Indian Oceans. Most have been illegally taken from the South Pacific.

In the early days of the commission, Japanese authorities provided a telling example. *Lung Yuin* was a Taiwanese-owned cargo ship flying a Panamanian flag and carrying frozen bigeye tuna to Japan. When authorities inspected, they found the tuna had been caught by 25 Taiwanese vessels and three Vanuatu-flagged fishing boats owned by Taiwanese companies. All 28 boats had given false information about where they had caught the fish. *Lung Yuin* had two logbooks – one true, the other false.

Another ship was discovered to be hiding an excessively large Atlantic bigeye catch, which had been relabelled to the Pacific. The Japanese authorities noted: 'This sort of organised laundering activity is ... widely conducted not only in the Pacific but also in the Atlantic and Indian Oceans.' They claimed around 18,000 tonnes of illegally caught bigeye tuna had been laundered in this way.[4]

While prevention of illegal fishing is a useful cover story for operations like that of the *Carl Vinson*, in reality the biggest danger in the Pacific is not from pirate fishing but from legal fishing – with China the biggest player by far. This country of almost 1.4 billion people has largely destroyed its own fisheries in the China seas and, like Spanish and European fishers who have lost much of the Atlantic and the North Sea fisheries, its fishing fleets need to press into new areas of the world. The Pacific, like the Southern Ocean, has the look of virgin territory.

The Pacific Islands Forum Fisheries Agency, which is based in the Solomon Islands and controls the exclusive economic zones for Pacific states, reported in July 2012 that in two years there had been a 125 percent increase in the size of China's South Pacific tuna fleet. In 2010 the country had 107 boats longline fishing for bigeye and yellowfin tuna. Now it had 241, ahead of Taiwan's 221. The extra boats had not been moved from other oceans: they were all new. And some of the Taiwanese boats were operated by Chinese companies. Many vessels flagged to the Solomon Islands, Vanuatu and Kiribati were also owned by Chinese companies.

A 2013 European Union Commission report 'The Role of China in World Fisheries' revealed that China's might is being turned on the global fishery. The country is now the world's largest 'producer' of fish – through wild fishing, aquaculture, and processing the fish of other nations (such as New Zealand) – and the export leader, contributing about ten percent by weight and 13 percent by value of the global trade in fishery products. Its annual fish exports, already worth around US$13.2 billion, are growing by 14 percent a year and go mainly to Japan, Europe, the United States and South Korea.

China's deep-water fleet, which depends heavily on subsidies to survive, is the largest in the world. It is bringing new technology and improved logistics to fishing. Unfortunately, as the EU report noted, what is not improving is the country's 'tendency toward secrecy in fisheries data, and near-complete disregard for public accountability of the use of public fisheries resources'.[5] Because China is not telling the world exactly how much fish it is taking, a lot of the scientific work needed to determine sustainability is unreliable. To add to the problem, the country does little to track bycatch. As a result there are large holes in international understanding of its fishing effort.

In 2012 the United States–China Economic and Security Review Commission, a congressional body, held public hearings and produced a report called 'China's global quest for resources and implications for the United States'. Lyle J. Goldstein, an associate professor at the US Naval War College, told the commission that China's status as the world's largest fishing power could be positive: if the country accepted and practised global fisheries norms, the world would see a 'giant step forward for environmental protection of the oceans in the coming century'.[6]

This was a diplomatic way of saying that China was raping the oceans, and increasingly in waters far from home.

Goldstein related how in 1985 the country had entered the realm of fishing powers by launching its first fleet of long-distance vessels. Within 15 years this had led to an almost fivefold increase in its catch.

Until the middle of the 1990s, when the environmental movement began to look more closely at seas beyond the North Atlantic, conservation was completely ignored. The result was horrifying. A Chinese fisheries expert, Mu Yongtong, would note in 2006 that most – if not all – of China's own fisheries had been fully exploited, and many were already exhausted.

In testimony to the commission, Tabitha Grace Mallory, a China specialist at the Johns Hopkins School of Advanced International Studies, had noted that while many in the West were calling this the 'China century', the Chinese were labelling it the 'ocean century'. With 85 percent of global fisheries fully exploited, overexploited or depleted, China's behaviour in international fisheries had considerable implications. Would the Chinese government be able to control the behaviour of its fishing companies and agents? Was the country abiding by the agreements it had signed? Would it be a responsible player in the global system?

China's deep-water fishing industry began when one of the country's state-owned fishing enterprises expanded outward to seek profit in Africa. Fishing came to be seen as an important part of the country's official 'going out' strategy, as elaborated in its 2001–2005 five-year plan, and Chinese companies were encouraged to search for new markets and invest abroad. Then, in 2010, a government task force called for distant-water fishing to be expanded for food security reasons.

As Mallory noted, China plans to expand its 2,000-strong deep-water fleet to 2,300 by 2015. This is seen as a way of both guarding China's oceanic interests and seeking international space for development: the more international space China has, the more resources and benefits it can obtain.

The industry plans to move into non-traditional fisheries such as Antarctic krill, a tiny crustacean unnecessary for human survival but a vital food for much of the world's marine life, including whales, penguins, seals, squid and fish. In 2010, China's Ministry of Agriculture carried out an inaugural exploratory catch of krill that brought in 1,846 tonnes.

Central government in Beijing and all provincial governments assist the fishing industry via tax relief, reduced import duties, reduced value-added taxes, reduced taxes on importation of second-hand equipment such as ultra-low-temperature trawling boats and purse seiner tuna boats, subsidies for boat renovation, subsidies for development and exploration of new fisheries, and fuel subsidies. Just as subsidised EU fishing wrought havoc in Europe and the North Atlantic, Chinese fishing seems set to do the same in the Pacific and Southern Oceans.

In 2012 I went to a meeting, in a downtown Auckland hotel, sponsored by a semi-governmental fund that usually pays for dance groups to tour, or the odd journalist to go and have a look around Tonga in the midst of democratic change. This time the Pacific Cooperation Foundation had flown in a group of fishing company officials from Pacific states, all looking for ways to make more money from the industry.

Despite countless political statements from practically every Pacific prime minister attending countless summits over decades urging the small Pacific states to cash in on tuna, few have managed it; most merely collect the relatively modest licence fees paid by the deep-water-fishing nations. There are no jobs on boats for the locals, and little service work ashore. Even Fiji, which has a tuna cannery and a fishing port, makes only a small return from Fiji waters itself.

A veteran operator, Russell Dunham of Fiji Fish Marketing Group Ltd, showed up. He had the look of a man who had seen and heard it all before. 'We all worry about the sheer number of vessels, and their technology means they now have double the firepower of five years ago in terms of how many hooks they can set,' he told me. 'It has become constant – not only are they here but there are many more on the way.'[7]

He complained that no one would talk about what China was doing, and nor would they say anything to Beijing. 'It is the same with the Fiji government, same with Pacific governments, same with the New Zealand government – no one is game to say anything.'

Chinese boats are built with 0.5 percent loans, many of them designed to keep shipyards working, and all receive about $300,000 a year as fuel subsidy.[8] The economics of tuna fishing meant they could not make money without the subsidies, Dunham said. For the Chinese government, it was not about making money. 'It's about positioning, getting as many boats into the Pacific as possible. When the time comes for quotas and slicing up the cake, China will be able to say it has 400 boats here.'

A subsidised fleet could keep fishing long after a normal commercial fleet would have to give up. 'The Chinese will fish until there is one tuna left in the ocean, and since the government is paying the bills the fish won't stand a chance,' Dunham added.

He believed the Western and Central Pacific Fisheries Commission would ultimately determine quotas by catch histories. If the Chinese can show they have had the most boats fishing for tuna, they will be entitled to the greatest number of quotas. 'That is what they want,' Dunham said, 'total control. Someone has to tell China to be a responsible fishing nation.'

Since the 1990s most large New Zealand fishing companies have had their fish processed in China. Smaller companies have retained New Zealand processing in a bid to create a point of difference for their product, but with limited success. As the closure in October 2013 of United Fisheries' Christchurch plant showed, it can be cheaper to send the jobs – 200 in United's case – to China. Christina Stringer found many companies believed this was the only way to remain competitive. 'China's processing factories are viewed as sophisticated operations with high hygiene standards,' she wrote in a detailed insider view of the fishing industry.[9]

Today nearly 95 percent of hoki caught in New Zealand waters goes to China for processing, much of it ultimately destined for McDonald's Filet-O-Fish burgers in the United States. Consumers, if they ask, are told they are getting New Zealand fish; China is not mentioned.

Stringer found that only one company in the New Zealand fishing industry, out of seven studied, was concerned about the impact that processing in China might have on the New Zealand brand. In her interviews she discovered that, when it came to fishing, the brand didn't really exist, although 'some buyers continue to differentiate in their purchasing requirements by requesting fish processed in New Zealand'.

China's extensive seafood reprocessing industry has competitive advantages: low-cost labour, and processing skills that guarantee more flesh taken from each individual fish. The factories are situated in the major port cities, with the largest in Quigdao, Shandong province. Workers in large factories earn around 1,000 to 1,500 yuan (US$160–240) a month, although wages vary widely across the industry.

Much of the seafood is processed manually, without mechanised filleting. Hand filleting and trimming mean the fish does not have to be so defrosted and fillets look better. Chinese plants claim they obtain fillet yields of between 68 and 71 percent of the fish, as against the European Union's 57 percent. Their workers are better than anybody at getting out the small pin bones consumers so dislike; they do it tediously, by hand. Salmon, just one example, has 36 pin bones.

Although China looks cheap, the drive to cut costs is sending processors to countries with even lower wages. According to Oceana, which was set up in 2001 and is now the largest international body working on ocean conservation, food safety officials believe the processing in some of these countries, including Bangladesh and Burma, ignores food safety standards. People in the United States have allegedly picked up cholera from 'illegally smuggled, temperature-abused' seafood imports; in Hong Kong, the government warns people against buying food from illegal seafood vendors because of the risk of cholera.[10]

Seafood such as shrimp can be high risk. A 2012 study by the Solidarity Centre, a Washington, D.C.-based body working to improve labour conditions around the world, quoted a shrimp-processing worker in Chittagong, Bangladesh, saying, 'Only when foreign buyers come to the factory are we issued boots and gloves, and as soon as they are gone, these are taken away again.'[11]

It is not just physical labour that produces processed fish in Chinese factories. Suspicions have long been raised over the way seafood is manipulated with the use of sodium tripolyphosphate (STPP), a chemical that is found in most industrial and household detergents. Fish are soaked in the chemical overnight to maintain taste and freshness. It also stops 'thaw drip', the liquid that oozes out when a frozen product defrosts, and increases the product's weight, and therefore its price.

After complaints from the European Union, China finally stopped using the chemical on fish bound for EU countries. The United States did not follow suit. Although its Environmental Protection Agency regards STPP as a pesticide, its Food and Drug Administration claims the chemical is 'generally recognised as safe' as a food preservative. Even if this is true, STPP is part of a consumer fraud: its only purpose is to hide the fact the seafood has been frozen, partially thawed, and then refrozen – after an increase in weight.

The almost fanatical Chinese demand for medicines from wildlife is replicated when it comes to fish. One target is totoaba, an endangered fish found in the Gulf of California. Many in China believe that consuming this fish's bladder promotes fertility, and are prepared to pay a high price for it.

Totoaba's desirability comes from its reputed similarity to bahaba, a fish found in Chinese waters that has been almost wiped out by overfishing. Today a single bahaba sells for the equivalent of US$500,000.

To destroy the stock of a species of fish simply for one small part of the creature's body is crazy enough, but it didn't stop with bahaba. Manta rays are taken for their gill rakers, which are added to medicinal soup and worth around US$250 a kilogram. Initially, most of the manta rays caught – around 5,000 a year – came from southern China. However, high demand has sent Chinese buyers to places such as Mozambique. In 2013 *The Guardian* reported that at one of Mozambique's top diving spots, Inhambane, the number of rays had dropped by 87 percent over a ten-year period.[12]

There is no scientific evidence that these products are any better than standard synthetic medicines that do not require the destruction of species.

And then there are shark fins. The lunacy of destroying a fish to make a soup stock out of just its fin is self-evident, but the Chinese, uniquely, continue to do it. Shark fin soup was once an élitist, strictly southern coastal Chinese dish. Nowadays, every Chinese person wanting to put on a bit of a show feels the need to order it. Asian crime gangs originally dominated the shark fin trade around the world until, sadly, the trade became mainstream.

I have never eaten shark fin soup and my experience with sharks has been limited to the occasional underwater meeting in Samoa and Niue; it's fair to say I did not linger. In 2013, however, fins entered my reporting world. Military-ruled Fiji has long been my beat, but when Voreqe Bainimarama – who seized power with a coup in 2006 and has failed to hold democratic elections since – got together with Euro-French company Airbus, it was not immediately obvious that sharks would be involved.

Fiji's airline, Air Pacific, was running a regular service from Nadi in Fiji to Hong Kong, although the number of tourists using it was limited. For sums not disclosed, Airbus told the Fiji regime it was to begin flying Airbus A330 aircraft – under a new name, Fiji Air – to Hong Kong each week.

The explanation as to why Fiji Air was flying to Hong Kong so often came in a striking article in the *South China Morning Post* that May. A Hong Kong couple named 'Cloudy' and 'Stephen' had won a honeymoon in Fiji by holding their wedding banquet without shark fin soup. The competition, run by the Hong Kong Shark Foundation under the tagline 'Happy Hearts Love Sharks', aimed to encourage bridal couples to set an example by eschewing the dish blamed for encouraging shark finning and depleting shark populations.

'We believe in love, and in the balance between nature and humans,' Cloudy was quoted as saying. 'I hope the competition will be held annually to educate people and change the unsustainable and cruel practice of shark-finning forever.'[13]

The irony was that, unbeknown to the happy couple, Fiji Air had flown them to Nadi on a plane that, on its outward leg, would be used to fly shark fins into Hong Kong.

The *South China Morning Post* reported that a letter signed by 78 organisations had claimed a 'substantial amount' of the shark fins imported to Hong Kong came by Fiji Air. Suspicions had been triggered by a speech given by Anthony Cheung Bing-leung, Hong Kong's secretary for transport and housing, when one of Fiji Air's new Airbuses passed through. 'There were only 45 tonnes of cargo being carried between Hong Kong and Fiji in 2009,' Cheung said. 'By the end of last year, the cargo volume was close to 1,000 tonnes.'

'Thanks to the close aviation links, we in Hong Kong can now enjoy various kinds of seafood products from the South Pacific, as Fiji is one of the major exporters of fish and fishery products to Hong Kong.'[14]

Alex Hofford of the Hong Kong Shark Foundation had said of the 20-fold leap in airfreight tonnage from Fiji to Hong Kong in just three years, 'It's not pineapples or electronics that are being flown here – you can be sure of that.'

He then learnt from a group of pilots that Fiji Air's new Airbus was 'basically a thinly disguised freighter' carrying shark fins harvested from the Cook Islands and Vanuatu. 'You may be on an Air Pacific flight where you think, this can't be making money – the plane is empty. But the fact is it's full of cargo. They can afford to lose money on the passenger side because they're making money on airfreight.'[15]

When I got involved in the story, I took up the issue with Fiji Air's spin doctor, Shane Hussein, a former Fiji Television frontman. He engaged in abuse and denial, attacking me personally and claiming I was trying to destroy the airline. Inside two months, after an 'investigation', Fiji Air announced it would fly only 'sustainably harvested' shark fins in future. There is no known way of sustainably harvesting shark fins. Fiji Air, which had spent millions on new aircraft and a new livery, had been caught out helping in the crooked business of looting the oceans.

Killing sharks sets off a process of unintended consequences. As the numbers of North Atlantic sharks have reduced, their prey have ballooned in number. These prey eat coastal bivalves such as mussels, scallops, clams and oysters, so these shellfish have become increasingly scarce. One result was the closure of a North Carolina scallop fishery that had flourished for a century.[16] Another was the escalating cost of clams, and therefore of clam chowder. Fewer and fewer restaurants in the America's northeast now serve this once hugely popular traditional dish.

19 The Russians are coming

As I mentioned earlier, the digital revolution has provided tools to protect the oceans. Transparent and easily accessible internet systems – available even on smartphones – can track merchant vessels. Just Google a ship and you'll usually be able to find out where it is located. But there has been a loophole for fishing boats. While a boat will often be loaded with electronics to find fish, once out at sea the captain can switch off the position-indicating system.

While governments and regional organisations ponder how to make fishing boats turn on, and leave on, their automatic identification systems (AIS), independent groups have stepped into the breach. One of the most striking is SkyTruth, which operates out of an office behind a draper's shop in Shepherdstown, West Virginia, population 1,736.

With an army of computer programmes, SkyTruth's volunteers use satellite radar imagery and data transmitted from ships to detect fishing boats that turn off or do not use AIS.

'Most people would be gobsmacked by what we are finding,' John Amos, the group's founder, told me. Many vessels were 'fishing-the-line' – catching valuable species right on the edge of other countries' exclusive economic zones. Using its satellite technology and smart thinking, SkyTruth had exposed alarming fishing practices right up to the edge of New Zealand's exclusive economic zone. A 'Spanish armada' of sophisticated fishing boats was forming 'picket fences' along the edge of the zone in order to nab bluefin tuna before they reached New Zealand waters.

In Chilean waters, too, SkyTruth was seeing Chinese-, Ukrainian- and Spanish-flagged vessels fishing right up to the line. 'With this concentration … you wonder how many of the migratory fish even make it through,' Amos told me.[1]

I came across SkyTruth as a result of an unusual trip to Whangarei, north of Auckland, one Saturday in November 2013. I had been tipped off about two odd fishing vessels at the port. The European Union-subsidised, Spanish-flagged ship *Carmen Tere* was on the slipway of Ship Repair NZ Ltd. Its Portuguese-flagged sister *Artico* was at a nearby wharf, guarded by a silent, shirtless, barrel-chested man.

Some of the Indonesian crew had asked to meet with people studying the fishing industry. Quietly, eight of the men came off *Carmen Tere* and were driven to a nearby motel.

'We know we are the poorest paid of all the fishermen in the world,' one told me. It was, by now, a familiar story. The men were earning US$325 a month – below the New Zealand minimum wage – and getting trivial catch bonuses. Around 30 Indonesians worked on the two boats; many had been on one or the other for over 60 days straight. Even within walking distance of supermarkets and fast food, they were being fed the same meal each day: fish bait – frozen pieces of mackerel and squid.

I asked through an interpreter whether they liked fish bait. The response was one of disbelief. They laughed politely and said the officers did not eat it. 'They eat Spanish food, meat, fresh fish, pasta.' Sometimes, they said, they would get the leftovers.

That was not the only deprivation they were suffering. They were not allowed clean fresh drinking water; their water came from a rusty tank, while the officers drank bottled water. And they had to wash 'out of the sea'.[2]

Carmen Tere and *Artico*, owned by Angelsonia Pesca SL of Lugo, Spain and part of the European fleet that had headed into the South Pacific in the wake of the collapse of the North Atlantic and Mediterranean fisheries, were licensed by their flag state to take swordfish and tuna outside the New Zealand exclusive economic zone.

The men said the ships were illegally taking Pacific bluefin tuna and showed us photos on their mobile phones of two large bluefins, each around 200 kilograms. They believed they were near New Zealand at the time; they had seen New Zealand-flagged vessels, and twice they had seen helicopters. The bluefin had been separately processed and then stored where it could be easily dumped. 'If the patrol comes, we dump it,' one of the men said.[3]

The men told me that three months earlier the ships had been boarded on the high seas by the New Zealand navy, a fact that was later confirmed for me by Scott Gallacher of the Ministry for Primary Industries. The inspector who boarded the ship did not talk to the crew. The bluefin tuna on board was not found. When the ships reached Auckland their deliberately mislabelled secret cargo had been transferred to a container and sent to Spain.

Most of the ships' crews were in debt bondage, having had to borrow the 4.5 million rupiahs (US$400) the agent demanded. Three had had serious accidents on their last voyage and had permanent numbness to arms and hands. When they had asked for medical treatment, the captain had just laughed, they said. One man had gone for 25 days before his wound was treated.

Trips took the men out into the Pacific for 60 to 80 days at a time. The calls back to Papeete and Auckland were brief. 'Never seen a girl for two years,' one man told me.

I phoned Angelsonia Pesca in Lugo. The man who answered the phone said it was untrue the crews were underpaid. He seemed surprised that we had copies of their contracts. 'The ship is a Spanish flag,' he said, when I suggested the crews were being paid well below New Zealand wages.[4] New Zealand rules, he said, did not apply.

When I wrote this story, I noted that *Carmen Tere* had also fished near New Zealand's Kermadec Islands. When the story appeared I was contacted by a woman named Bronwen Golder on behalf of the Kermadec Initiative, a project backed by the rich United States-based Pew Charitable Trusts. The Initiative was campaigning for the seas around the Kermadecs to be made a marine sanctuary.

Golder had information on a 34-year-old Cambodian-flagged tuna boat, *Gral*, which had left Suva and made straight for Raoul Island, the largest island in the Kermadecs, despite having no licence or permissions and not complying with normally strict quarantine procedures. The information about the boat had been obtained by SkyTruth, which Pew had commissioned to monitor what was happening around the Kermadecs.

They had seen *Gral* pull into Suva on January 17. A month earlier the ship, owned by Aururoa Shipping of South Korea, had been called *Arche*, and before that *Gaia*, *Oryong 332* and *Sam Song 507*. On January 20 it had sailed out of Suva and headed to Raoul, arriving at 2.55 a.m. on January 23. It had passed none of the mandatory checks for vessels heading to the Kermadecs.

A Tauranga ship, *Claymore II*, under charter to the Department of Conservation, had found *Gral* anchored in a bay. It was never discovered what the ship was doing unsupervised at one of the world's ecological treasures.

Soon after this, SkyTruth data showed the world's largest fishing factory-freezer ship and its fleet of catcher trawlers passing through the Kermadec Islands and New Zealand's exclusive economic zone. Under international law, ships of any kind are allowed to make innocent passage through the exclusive economic zones and territorial waters of any nation. 'Innocent passage' is a broad term. It could, for example, allow a nuclear-armed warship to pass through New Zealand's Cook Strait despite the ship's presence being illegal under New Zealand's nuclear-free legislation. Nevertheless, the presence of this ship was alarming.

A Chinese-owned 49,367-gross-tonne converted oil tanker flying a Russian flag, *Lafayette* uses giant hoses to suck the catches from its attendant trawlers. Its target on the high seas was South Pacific jack mackerel, used as feed for farmed salmon. *Lafayette* had its AIS turned on, but the six trawlers working with it, all Peruvian-flagged, did not.

Lafayette is owned by a Hong Kong-based, Bermuda-registered company Pacific Andes International, which in 2008 spent US$100 million remodelling the former tanker. The only Russian-flagged ship fishing in the South Pacific, in 2010 it reportedly took 41,315 tonnes of mackerel east of New Zealand, an increase of 136 percent over the previous year.

Other nations accused Russia of faking its statistics in order to later get a better-sized quota from the South Pacific Regional Fisheries Management Organisation, SPRFMO. This multinational body tries weakly and almost desperately to prevent the total pillaging of the region. After it met in Auckland following *Lafayette*'s first Pacific tour, it had stripped Russia of its quota.

After Russia appealed, I obtained documents that showed several countries had doubts about *Lafayette*. China described its operations as 'dubious'. Chile said it was 'seriously undermining trust and confidence'. The European Union had demanded an investigation after French officials found the ship's owners had lied about its fishing capabilities.[5]

In 2009 Russia had advised the names of the vessels it would send into the South Pacific. *Lafayette* was to fish for 'horse mackerel'. Meanwhile, SPRFMO read a news story that said *Lafayette* was a mother ship or processing vessel, working with trawlers flying Peruvian flags. The organisation wrote to Moscow asking if it was a mid-water trawler or a fish-processing ship. Russia said it was a trawler actively fishing.

When *Lafayette* called into Papeete in Tahiti in January 2010, French authorities inspected and found no fishing gear or fishing equipment on board. Photos clearly showed the vessel has never fished, and its logs showed it had never been in the area where it claimed to have fished.

A South American group, Centro Desarrollo y Pesca Sustentable, claimed what was happening was a 'race for overreporting'. *Lafayette*'s catch was 'a fiction'.[6] France sent Moscow a diplomatic note saying it believed *Lafayette* was not fishing but was a processing vessel, and China expressed concern about the legitimacy of its catch figures.

In response, Russia merely complained about the French inspection of the *Lafayette*. Meetings followed and a scientific group reported concern at the possible double-counting of Russian and Peruvian-reported catches.

With *Lafayette* about to be banned from the region, Russia complained it was being obliged to provide far more information about the high seas than it was used to giving. 'It's a major burden, and there's a lot of information to process,' a Russian representative said.

Under the rules of the convention, Russia was able to appeal to a special panel sitting in The Hague. Here, legal academics led by an American and supported by a Russian political adviser and a Chilean heard the case. The transcript showed that, while they considered the issue in depth, they also worried a bit about the weather. The Russians won their appeal, but what was perhaps more intriguing was that the process gave the public – or at least those interested enough – a wealth of background material into how fishing nations see the high seas. In summary, their view appeared to be one of exploitation rather than conservation. The victory before the tribunal was a legal not a moral one.[7]

Early in 2014 *Lafayette* came back. After passing through New Zealand waters it moved out into the South Pacific with a fleet of trawlers ready to hoover up tonnes of mackerel.

20 Wars

W hile the fishing business has grown exponentially and alarmingly in the 21st century, conflict between nations over who has the right to fish what and where is nothing new. There have even been wars between allies over fish. By good luck rather than good management, there have to date been no fatalities.

While most of these fish wars have gone largely unnoticed, one received a lot of attention if only because of its oddness – namely, the skirmish between Iceland and Britain over cod. The widespread destruction of the North Atlantic cod fishery led these countries into three clashes. The first came in 1958, when Britain found it could not prevent Iceland extending its fishing limit from four miles to twelve. The Royal Navy went in to protect British fishing vessels, and at one point shots were fired across various bows. There was annoyance in the British parliament when it was discovered the navy's fuel bill for the exercise easily exceeded the proceeds of fish caught. But then Europe has always had a rich tradition of subsidising, both with cash and political action, its fishing industry.

In 1972 Iceland extended its fishing limit to 50 miles. An Icelandic patrol boat tried to chase 16 foreign trawlers out of its waters. One, a British boat called *Peter Scott*, refused to identify itself and instead played 'Rule Britannia' over the radio. The Icelandic coast guard cut its net. Another

British trawler, *C. S. Forester*, was hit with a couple of shell rounds that damaged its engine room and water tanks. The ship was boarded, taken to Iceland, and eventually released after a fine was paid.

A brief settlement followed. Britain was allowed into the newly extended waters provided it took only 130,000 tons of fish a year. The agreement was good for two years. When it expired, Iceland extended its authority out to 200 miles, and in 1975 and 1976 Icelandic coast-guard vessels began cutting British nets, and ships were rammed at sea.

Iceland's six ships and two converted trawlers were up against 22 Royal Navy frigates and some support ships. At one point a British frigate came upon an Icelandic gunboat on the high seas. Happily, both captains recognised the issue for what it was worth. One captain used a loud hailer to fire across biblical verse, to which the other responded with more readings. Journalists present said the Icelanders were the winners of the scriptural contest.

The cod wars went to the United Nations, which did nothing. However, Iceland had an important card, the NATO base at Keflavik, which the United States found useful in those days of the Cold War. When Iceland threatened to close the base, the Americans quickly insisted on a peace agreement. Britain, which used to take around ten million kilograms of fish a year from Icelandic waters, saw hundreds of fishermen, as well as people in shore facilities, lose their jobs. Iceland was the victor.

In 1995, a lesser-known fish war lined up Canada, backed by Britain and Ireland, against Spain, backed by the European Union. Because stocks of cod had crashed, Canada had imposed a moratorium, and the country's fishermen had switched to catching turbot. Canada stopped a Spanish trawler, *Estai*, in international waters and arrested its crew, claiming they were illegally overfishing Greenland turbot on the Grand Banks, just outside Canada's exclusive economic zone.

A Canadian fisheries patrol vessel, *Cape Roger*, fired shots from a .50 calibre machine gun across the bow of *Estai* to stop the ship. Another Canadian ship, *Sir Wilfred Grenfell*, used water cannons to deter other Spanish fishing vessels. In response, the Spanish navy sent a patrol boat,

Atalaya, to protect its ships, and ordered frigates and tankers into the area before backing off. Canadian ships then cut the nets of another Spanish trawler.

Eventually Canada and the European Union negotiated a settlement. This was rejected by Spain. The EU prevailed on Spain to accept it. As with the South Pacific mackerel plunder, reaching deals on fisheries disputes can be Byzantium and impenetrable to the lay observer. In the end the parties agreed to an overall reduction in turbot catch, compensation to Spain for ship and net damage, and changes to Canadian law to allow a newly created international regime to monitor the fishery.

With the world's population slated to double by 2050, control of protein will become increasingly important. Conflict will come, even in the South Pacific. In 2011 Fiji and Tonga sent gunboats at each other, inconclusively, over who controlled the Minerva Reefs on the maritime border. The unpopulated, largely submerged atolls had been claimed by Tonga in 1972 after they were partially raised above sea level by an American group intent on establishing an independent nation. Although the Tongan claim was recognised by the South Pacific Forum, it has never been accepted by Fiji. In the face-off, Tonga won, for now.

In 2013 new tensions, based in part around fishing, were brewing in the South China Sea. Countries such as the Philippines were declaring fisheries regulations over areas of the sea that others, notably China, claimed as their territorial waters. Other countries, including Japan, Taiwan and Vietnam, were making it clear they would obey regulations only when it suited their foreign policies.

Through the United Nations and its Convention on the Law of the Sea, there are papal-style mechanisms in place to settle such disputes. It needs to be that way because the resources of the ocean are still, in essence, viewed as the common wealth of all people. Exclusive economic zones, as the South Koreans reminded New Zealand, do not give rights of ownership over their content. The zones imply local management and control, but if they are used to exclude others they can be contested.

South Korea's argument was that New Zealand was unable to fish its entire exclusive economic zone and so should allow others in. This was a long way from outright conflict, but it illustrates that in hard times, when fish are in great demand, there is potential for dangerous encounters on the high seas of the South Pacific and Southern Oceans. That New Zealand and South Korea are nominally friends, and that Wellington spent blood and money to help save Seoul from communism in the 1950s, is unlikely to cut much ice when it comes to access to food.

The State of World Fisheries and Aquaculture – SOFIA – is fishing's equivalent of the Domesday Book. Published by the Food and Agriculture Organization of the United Nations, it comes out every two years and summarises the world's fish stocks in oceans, lakes and rivers, and in aquaculture.

Like anything to do with the UN, the report's contents have often been tampered with by diplomats and manipulated by politicians. The Chinese repeatedly provide misleading statistics that suggest they are taking fewer fish from the oceans and growing more through aquaculture. Hence, when *SOFIA* opens with a comment expressing concern about the state of stocks exploited by 'marine capture fisheries' – the FAO's term for wild fishing – it is reliable confirmation that the world's fisheries are in trouble.

In reality nearly a third of all fish stocks are overexploited – that is, they are producing lower yields than their biological and ecological potential, and are in need of strict management plans to restore their full and sustainable productivity.

Every fish stock that can be economically caught is sliding toward disaster. 'The increasing trend in the percentage of overexploited, depleted and recovering stocks and the decreasing trend in underexploited and moderately exploited stocks give cause for concern,' the 2013 *SOFIA* stated bluntly.[1]

The solution it proposed was to rebuild overexploited stocks through putting in place effective management plans. This will not be easy. Most of the world's top ten species – which account for around 30 percent of all fish consumed – are regarded as fully exploited, with no potential for increased production.

These include the two main stocks of anchoveta (Peruvian anchovy) in the southeast Pacific, Alaska pollock in the north Pacific, and blue whiting in the northeast Atlantic. In addition, Atlantic herring stocks are fully exploited in both the northeast and northwest Atlantic. Japanese anchovy in the northwest Pacific and Chilean jack mackerel in the southeast are both considered to be overexploited. Chub mackerel stocks are fully exploited in the eastern and northwest Pacific.

Of the seven principal tuna species, one-third are estimated to be overexploited and 37.5 percent fully exploited. 'In the long term,' *SOFIA*'s report warns, 'because of the substantial demand for tuna and the significant overcapacity of tuna fishing fleets, the status of tuna stocks may deteriorate further if there is no improvement in their management.'

The population of large, predatory, open-ocean fish such as tuna, swordfish and marlin declined 90 percent between 1952 and 2005 and is still falling. No fate is quite as bad as that of Atlantic bluefin tuna. Bluefin fishing, led by Japan, is a monumental example of mismanagement and greed. The country has steadfastly ignored scientific findings that it is eating the species to extinction. In 2009 around one million bluefins were caught, out of a total population of only 3.75 million. Atlantic stocks have dropped 80 percent in only 20 years.

In 2010, efforts were made at the triennial meeting of CITES, the Convention on International Trade in Endangered Species of Wild Fauna and Flora, to get bluefin tuna listed as one of the world's most endangered species and introduce a ban on fishing it. It beggars belief that the Tokyo government, backed by the fishing industry, marshalled forces to argue the Japanese had a right to eat a species into extinction.[2]

Ignoring the drastic statistics, they argued their case on the basis of some kind of racial entitlement. 'This is like telling the United States to stop eating beef,' Kimio Amano, a 36-year-old broker at the Tsukiji fish market in Tokyo told *The Guardian*.[3] He and others claimed that bluefin tuna just needed to be better managed, and that a ban on one species would lead to bans on others.

Despite the fine words, what was behind the resistance was not culture but money: a single bluefin tuna can sell for US$100,000. The best specimens are not 'shipped' to Tokyo: they are flown there as fast as possible.

'Our biggest hope is that the [very high prices] don't spread to the Pacific,' Tadao Ban, head of the Tokyo cooperative of large fish dealers, said. 'For this reason we are promoting strict resource management. We are even supporting putting a tag on each and every tuna caught.'[4] Many would argue that such action would be too little and too late. Pacific and southern bluefin tuna are already showing signs of being in the same parlous position as Atlantic bluefin.

Regional fisheries bodies manage tuna, or try to. In the case of Atlantic bluefin, the responsibility rests with ICCAT, the International Commission for the Conservation of Atlantic Tunas. In 2009 ICCAT scientists reported that the spawning biomass of Atlantic bluefin was less than 15 percent of what it had been before industrial fishing began.

Although the bluefin market is almost exclusively in Japan, the money is such that many European countries, including France, Spain, Italy and Greece, fish for the lucrative catch. With subsidies from the European Union, these countries have built ever bigger boats to narrow down the hunt to the last bluefin. France is one of the worst offenders. Year on year, it has ignored its quota, in some years taking up to 60 percent more than it is entitled to.

Greenpeace International has been highly critical of ICCAT. Its 'failure to halt the decline of the east Atlantic bluefin population – despite warnings provided by the collapse of the west Atlantic population and the alarm bells rung both by the scientific community and NGOs – is a damning indictment of the organisation and of the failure of contracting parties to fulfil their responsibilities,' it said in 2007.[5]

Responding to international clamour, in 2010 the European Union's Fisheries Commission temporarily banned fishing for bluefin tuna in the

Mediterranean and eastern Atlantic by purse seiners. France and Greece appealed to the General Court of the EU, but the short-term ban was later upheld.[6] In 2014, ICCAT retained annual quotas of 13,400 tonnes for the eastern Atlantic and 1,750 tonnes for the western Atlantic, saying in its defence that Japan and other 'contracting parties' had pressed for the quotas to be increased by 400 tonnes.

Many people see the future as aquaculture. It is true that 'farmed' fish production has expanded almost 12-fold over three decades, at an average annual rate of 8.8 percent, to the point where it now produces more fish than does the wild marine catch. But aquaculture is far from foolproof. In recent times disease outbreaks have affected farmed Atlantic salmon in Chile, farmed oysters in Europe, and farmed marine shrimp in several countries in Asia, South America and Africa, resulting in partial, and sometimes total, loss of production. In 2010 aquaculture in China suffered production losses of 1.7 million tonnes caused by natural disasters, diseases and pollution. Disease virtually wiped out marine shrimp farming production in Mozambique in 2011.

A big problem with increasing aquaculture is the need to feed the fish. With salmon farming, for example, to get a kilo of farmed fish you have to feed in as much as four kilos of wild-caught fish. Lower value fish such as mackerel could be used, but ultimately this does not look sustainable.

According to the FAO, a third of all farmed fish production, mainly of bivalves and filter-feeding carps, is currently achieved without artificial feeding. But this kind of fish farming is declining as demand grows for bigger fish, more plentifully available.

Happily, bugs may provide a solution: in Chile and Norway, salmon farms have begun experimenting with breeding insects to feed fish, and in September 2013 America's National Public Radio reported on a small industrial company in Ohio that was breeding black soldier flies. These insects don't spread disease and seldom bother people. They mate and produce larvae. The larvae are insatiable eaters and will eat anything. The company, EnviroFlight, feeds them waste from an ethanol plant, but they could equally well be fed waste from a brewery or a slaughterhouse. After the larvae reach the right size, EnviroFlight cook and dry them. This produces a rich product, suitable for feeding to fish.

EnviroFlight's founder, Glen Courtright, reckons black soldier fly larvae could simultaneously solve two enormous global problems: the waste problem and the food supply problem. 'We have a protein deficit. We are heading for nine billion people on the planet. We don't know how we'll feed them.'[7]

Perhaps people will worry about what the expensive farmed salmon on their plate has been fed, but so far this has not been the case. And what they're eating now is in most instances endangering the world's marine environment.

21 Hope

n September 2009 something highly unusual happened: New Zealand made the front page of *The New York Times*. Under the headline 'From Deep Pacific, Ugly and Tasty, With a Catch', reporter William J. Broad revealed to readers that McDonald's Filet-O-Fish involved 'an ugly creature from the sunless depths of the Pacific, whose bounty, it seems, is not limitless'. The world's insatiable appetite for fish was heading towards hoki – or what he called whiptail. (It is also known as blue grenadier, blue hake and New Zealand whiptail.) McDonalds alone was taking nearly nine million kilograms a year from New Zealand. Hoki looked 'exceedingly unattractive', Broad said, but when turned into fillets its flesh became moist, slightly sweet and tasty.

While many people believed hoki fishing was sustainable, this was open to question, he said. 'Without formally acknowledging that hoki are being overfished, New Zealand has slashed the allowable catch in steps, from about 275,000 tons in 2000 and 2001 to about 100,000 tons in 2007 and 2008 – a decline of nearly two-thirds.'[1]

He compared the situation with orange roughy, where catches had fallen drastically in the early 1990s when New Zealand hit the stock with intensive fishing. As orange roughy stocks fell, hoki fishing rose. Hoki's shorter lifespan – up to 25 years, compared with orange roughy's 140 plus – and quicker pace of reproduction promised sustainable harvests. And its dense spawning aggregations from June to September made colossal hauls relatively easy.

From 1996 to 2001, when New Zealand's Ministry of Fisheries had hoki quotas of 275,000 tonnes a year, 'dozens of factory trawlers plied the deep waters, and dealers shipped frozen blocks and fillets of the fish around the globe,' Broad reported. But with the decline in stocks and cuts to quotas, some American fish re-sellers such as Yum! Brands Inc. – which owns or licenses KFC and Pizza Hut – decided not to sell hoki. Denny's, an American family restaurant chain with franchises as far afield as China, decided it would sell it only in New Zealand.

The *New York Times'* article plainly spooked people who give a second thought to the food they're eating. Fearing severe damage to the US$130-million-a-year hoki exports, the New Zealand Seafood Industry Council purchased Google ads to target people following the story. Google AdWords say almost nothing in themselves; rather they contain internet code, which on a click carries the reader to the advertiser's preferred website. At the time of the hoki drama, some of the Google AdWords said only: 'Hoki Facts'. Today the technology has developed further: words in a story on a page are enough to trigger ads aimed at the user.

The council was alarmed because hoki was being linked to McDonald's. 'That guaranteed the story went into every paper in America,' its spokes-person Sarah Crysell said. 'We needed to quickly correct a front-page article we thought was obviously flawed and contained significant and damaging distortions.'[2]

On *The New York Times'* website, the article had a link to the council's hoki page. The council quickly changed the page's contents to expose what it said were errors in the story and to provide scientific data. It completely rejected the article's claims. 'New Zealand takes pride in the fact that both its hoki stocks are among the best managed stocks in the world,' the site said.

Within 24 hours of the story appearing, 78,000 people had clicked *The New York Times'* link to the council's website. Usually only a handful of people a month looked at the website.

The council also hired a New York public relations agency, CounterPoint Strategies, to launch a digital counter-attack. People searching for the story – or for linked topics such as 'hoki', 'New Zealand' or 'fish' – found themselves transported to the council's hoki page. The agency went as far as to turn the reporter's name into an AdWord, to let him know 'we meant business and hold him responsible'.[3]

CounterPoint's Jim McCarthy told me in an interview that *The New York Times* story was 'heavy artillery' and the strategy was to limit the damage. It worked: traffic to the council's site doubled in three weeks. 'I don't know if we saved the species but we certainly taught *The New York Times* a lesson,' McCarthy said.[4]

Harvard University's Nieman Journalism Lab, in a study called 'New public relations: Beating back bad press with Google AdWords', hailed the campaign as 'a feat of public relations genius'.[5]

If there is a point in the McDonald's hoki story it is that, public relations strategies notwithstanding, educated consumers can sometimes end foul fishing practices – from massive bycatching of other species to brutal oppression of crew on fishing boats. The need to be educated about what is in our food chain has never been greater, especially when it comes to fish. The boats may sometimes be Dickensian, but they have access to technology that makes every living thing in the ocean a target. And fish are now often regarded as the beginning of the process, not the end product. A human being would find Antarctic krill virtually inedible, but subject it to industrial cropping and scientific transformation and it ends up as a capsule of magic oil for well-off people. Left alone, though, the krill forms the basis of a global marine ecosystem that, no matter how clever we are, we cannot replace with technology.

As Michael Lodge, now an independent adjudicator with the Marine Stewardship Council, put it in an interview with *Islands Business* magazine:

'The issue is not to try to hold back technology, but to ensure the rules take into account the greater fishing power of vessels compared to 20 years ago.'[6]

At first blush, the slave-like conditions endured by crew on many of the world's fishing boats is not something over which consumers would seem to have any say. Rather, we would expect our governments to do something about it. But, as shown in this book, successive New Zealand governments knew what was going on and did nothing.

As I write, the current New Zealand government is deciding whether to require foreign charter vessels operating in its country's waters to bear the New Zealand flag, meaning they would be answerable to New Zealand law. The parliamentary bill that would bring this about underwent some mysterious rewriting before emerging from a parliamentary select committee. The Ngāpuhi tribe had scored a sweetheart deal to allow it to continue using slave boats. The chair of its organisation Te Rūnanga-Ā-Iwi-O-Ngāpuhi, Sonny Tau, told me this almost gleefully. Without the slave boats, he claimed, New Zealand would lose $300 million. 'The fact is Korean fishers are the only fishers in the world who fish New Zealand squid quota,' he said. The quota is held entirely by Māori.

In Tau's opinion the bill had gone too far. All that was needed was for foreign charter vessels to be 'deemed' to be New Zealand boats while in New Zealand waters. That way they would be required to follow labour and human rights laws.

'We would never condone and have absolutely no tolerance for people who do not follow the rules,' he told me.[7] The truth was somewhat different: the iwi had taken no notice of the abuses at sea until others had made it an issue. They had gambled that the world would not notice the terrible treatment and conditions endured by men fishing for their quota. But if American consumers, followed by those in the rest of the world, wake up to the truth, New Zealand's *100% Pure* marketing campaign will come to mean something more akin to the dark side of *The Lord Of The Rings* than to beautiful beaches and healthy food.

Consumer knowledge is a powerful weapon. Most of the 38 million people who live in California do not want to be tainted with slavery. In 2010 the senate passed the California Transparency in Supply Chains Act. Since January 1, 2012, retail sellers and manufacturers doing business in California have been required to publicly disclose their efforts to eradicate slavery and human trafficking from their direct supply chains. The act affects over 3,000 companies, including major brands, retailers, vendors, and suppliers with headquarters outside California and even outside the US.

The same consumer power is already producing changes at sea. A lot of canned tuna around the world is now labelled 'FAD free' – a reference to harmful fish aggregation devices. In the open ocean, tuna congregate under floating objects such as palm trees, logs, weed mats, dead whales and half-sunken boats. Purse seiners take advantage of this by dropping into the ocean large manmade FADs, constructed of anything from simple wooden frames to high-technology carbon fibres and steel complete with position-transmitting devices. They leave them for several days before returning. The objects attract mostly juvenile bigeye and yellowfin tuna, which are then virtually vacuumed up by the purse seiners. Adult bigeye that have gathered are attacked by longline fishing boats.

These fish aggregation devices have opened up to exploitation a vast new area of the Pacific – the otherwise untouched tuna-rich waters between French Polynesia, Kiribati's Line Islands and the Galapagos Islands. Around the Philippines there are an estimated 4,000 fixed FADs, known there as payaos – one every 55 kilometres. The Bismarck Sea to the north of Papua New Guinea has hundreds of scattered FADs. But consumer momentum against these potentially harmful devices that can lead to overfishing, changes in fishes' natural movement patterns, and high volumes of bycatch, is building as the word spreads.[8]

Particularly in the world's richer economies, people who like to eat fish are increasingly demanding guarantees from retailers that the fish they have on sale is not only of high quality and safe to eat, but derives from fisheries that are sustainable, and where workers are paid a fair and reasonable wage.

One of the key tools is traceability. Just as technology can find wild fish stocks, it can trace and track a fish from the moment it swims into a net until a consumer buys it – or some of its parts – in a supermarket. Non-governmental groups such as the Marine Stewardship Council are now monitoring marine food chains worldwide.

The issues surrounding commercial fishing – human rights, fair wages and conditions, sustainability of the resource, environmentally responsible catch methods – come down to the shadowy world of fishing companies and to the individual fishing boats, their officers and the crews. These people and companies can be pirates or saints. They can be obliged to be one and not the other.

While skippers and officers work for the fishing companies, in law they are responsible for what happens on board their ships: this is why they are able to collect the larger share of the catch bonuses. Companies with an ask-no-questions approach allow questionable ethics to profit – for those fishing and for those who own the boat and ultimately the catch.

The FAO is slowly working through hurdles and bumps to create a global record of fishing vessels and cargo ships specifically designated for carrying fish cargoes. At the moment such a list does not exist: there are only lists, created by regional authorities, of boats allowed to go after particular catches such as toothfish. As we've seen, the same fishing boat can go through life under many different names and numerous flags. The renaming of fishing boats is almost never done for sentimental reasons. While renaming is not always dubious or criminal, it often is.

A global registry should not be hard to create: since 1987 every vessel over 100 tonnes has been required to have an International Maritime Organization seven-digit number. The IMO, a United Nations body, adopted this system to enhance 'maritime safety and pollution prevention, and to facilitate the prevention of maritime fraud'. The number is granted as the ship is built, and is never changed despite changes in owner, registry or name.

The numbers are assigned by what is now called *IHS Fairplay*. IHS, a US-based business information service, took over *Fairplay* magazine from Lloyd's Register.

Fairplay was founded in 1883 by Thomas Hope Robinson, a merchant seaman. When Robinson finally came ashore, he created what would become international shipping's weekly bible. In the first issue he wrote, 'There is so little fairplay in the world. If our own efforts succeed, we shall have taken the first steps towards promoting the habit of calling things by their right name and looking at them through uncoloured spectacles.'[9] It is fitting that something he created should be used to help remodel the world fishing industry.

That industry is in a grim state, as is inevitable for any industry that bases a significant part of its operation on poor wages. Its practices need greater light shone on them. As consumers, we need to look over the horizon in a way we have never done before, and question what is put on our plates.

Right on the deadline for submitting this book to my publisher, a series of emails arrived. My work on this project had run to nearly three years and I was beginning to get a sense – or perhaps just a desperate hope – that progress was happening in the fishing business, at least when it came to the brutality of man to man. Then I read the emails.

Someone on board *Shin Ji* – the hagfish trawler that had escaped bankruptcy in Auckland by moving to Suva – reported that a 28-year-old, Japanese-flagged tuna longliner, *Chokyu Maru 78*, had entered Suva harbour. On board was the deep-frozen body of a 33-year-old Indonesian man, Untung Subejo.

Subejo had gone stir crazy on *Chokyu Maru 78* when it was off Nuku Hiva in French Polynesia, fishing for tuna. It was said he had been at sea for several years, having been transhipped between vessels.

Near Nuku Hiva he had tried to kill himself by jumping overboard. He was hauled back. He had then tried to persuade the captain to end his modest contract and let him off the ship. Instead, he was strapped up and put into a bathroom-toilet near the bridge. For some reason, the door was then welded shut.

The ship set sail for Fiji. Four days later Subejo's battered and bruised body was found inside. He was dead.

Arriving in Suva, the captain asked the authorities whether the ship could drop off the body and go. Apparently the Fiji authorities weren't too bothered by the death – it was just seen as the price of Pacific tuna. The 20 remaining Indonesians aboard wanted to leave the ship. The Indonesian hiring agent made it plain they would each have to pay him US$2,000 if they did so. The men would have been earning no more than US$250 a month.

I end this with a somewhat more romantic story of the South Pacific. Parrotfish belong to a family of around 90 coral-living subspecies. I have mixed feelings about parrotfish. When I first snorkelled on the reef off my Samoan home I was struck by their colour and beauty, and locals said they were good to eat. I tried to spear one once but I was never much of a shot. Years later, the parrotfish exacted a kind of revenge when I ate one and nearly died.

Like many fish that live on reefs or close to shore in tropical and subtropical waters, parrotfish can become tainted with toxins that cause the illness ciguatera. There is no easy way to detect these toxins, which are produced by certain algae. Eat tainted fish, as I did, and you are in big trouble. Ciguatoxin is a cumulative toxin that remains in the body forever, so were I to repeat that Fiji meal I would become very unwell.

The presence of the ciguatera-causing toxins in reef fish is usually related to environmental damage; either a storm has damaged a coral reef or onshore pollution has affected it. Ciguatera used to be common in fish around Mururoa Atoll in French Polynesia, where the French had carried out atmospheric nuclear tests from 1966 to 1974.

The toxin has long afflicted Pacific communities. In some places, such as atolls in Kiribati, the old people eat fish first as it is believed they will be the most sensitive to the toxin and can protect the others.

A Cook Islands researcher, Teina Rongo, has suggested ciguatera may even have been the motive behind Polynesian voyaging. When the toxin broke out in the food chain, famine would threaten: the islanders, with their main source of protein inedible, would have to find somewhere else to live and fish.

Parrotfish, which can get up to 1.3 metres, rasp algae from coral and also eat the coral itself. What goes in through a parrotfish, as is the way of nature, comes out as mostly white sand. A New Zealand environmentalist and blogger William Hughes-Games did the numbers and worked out that the average parrotfish produces 90 kilograms of sand a year; a thousand parrotfish will produce 90 tonnes.

For people living on low-lying atolls, such as tens of thousands of Pacific Islanders, this is life-saving: parrotfish excreta adds to their landmass. But the fish, although it has little commercial importance, is still regularly fished in the tropics: it is easy to enjoy a fresh fish dinner rather than ponder why your atoll is slowly washing away.

The contribution of the parrotfish is becoming ever more important in the context of global warming and sea level rise. Atolls are fragile but they have survived previous sea level rises and falls; the coral reef top is usually always exposed at low tide, with the atoll itself rising up to five metres above high water behind them. The reason for that is the sand the parrotfish makes in its routine day.

Sand does not move much over time, and what the parrotfish produces is enough to retain basic life and the durable coconut palm. International conventions and carbon trading may do something globally to mitigate the impact of sea level rise, but one strong local action will make a big difference at home: leaving the parrotfish alone.

Hughes-Games has set out the basic principles to save atolls: do not damage the corals; never kill a parrotfish; leave rabbitfish (28 species of the family *Siganidae*) alone too. The sand production of rabbitfish is not large, but they clean coral and keep the reefs healthy. 'Short of a global situation that kills the corals, the fate of the atolls is in the hands of the local people,' Hughes-Games says.

It is not the first time the Pacific has provided a powerful philosophic message – from the noble savage, to the earthly paradise, right through to the environmental ruin of Rapa Nui, Easter Island. The story of the parrotfish, though, may be the most profound of all: eat all the parrotfish and the land upon which people live will disappear. Leave them alone and generations will thrive. If we respect and care for the sea, the creatures who live in it and the people who sail upon it, we will all have a better world.

Epilogue

The book was almost complete but there wasn't an obvious ending: many of the injustices I had spent a couple of years writing about were continuing. For a reporter this would have been a case of just getting on with the story, covering any new developments as they arose, but for the author of a book the lack of a neat finale poses a problem.

It was early April 2014 when I phoned Hasan Nurhasan. The timing had to be precise thanks to time zones. As I was heading to a deadline in New Zealand, Hasan was attending morning prayers at his local mosque in Tegal, Central Java, a province of Indonesia. He would then have a free hour or so, but he was a school teacher and much of his day would be devoted to children while mine quickly faded into night.

Tegal is where many of the men who crew the fishing boats around New Zealand come from. The city, with a population of about a quarter of a million, is crowded and poor, with high unemployment. Its young men hold hopes about crewing fishing boats around the world. Hasan speaks and writes English, and as a local teacher he helps them negotiate the difficult world of signing up to fishing companies.

I'd phoned him because I was looking for some kind of happy ending, something to show for my reportage other than stories in the newspaper and a disorganised box of paperwork. I had heard from my sources that crews off the Korean ships I had been covering had linked up with various individuals and organisations – many of whom still do not wish to be named and prefer to work under the radar – who had started legal processes to win justice and compensation for the men for the abuses they had suffered. Four hundred and eighty Indonesian men were seeking US$25 million in wages they had not been paid while fishing for New Zealand companies.

'The crews don't expect as much as that,' Hasan told me when we finally got to speak. 'They just want the balance of what they worked for.'

Justice was not going to easy to come by. The agents who had recruited the men had turned up in Tegal's kampungs with a delicate mix of threats, hope and promise. They were representing Indah Megah Sari in Jakarta, the company that organised low-paid crew for the *Dong Won* boats that worked on charter for Sanford Ltd.

Indah Megah Sari had placed advertisements in the classified sections of local newspapers. The advertisements, when translated into English, read: 'To all crew… who worked in Korea for Dong Won Fisheries Company F/V *Dong Won 701*, *530*, *519* who worked since June 2009 until May 2012. Can take the rest of salaries caused by the miscounting by Dong Won Fisheries Company.'

Under the deal, Indah Megah Sari would get the crewmen to dump their legal representatives in New Zealand in return for a 'peace agreement'. Hasan told me the money the men were being offered was much less than they were entitled to. In the case of the Korean boat *Pacinui*, owned by Juham Industries and working in New Zealand, the men would also be guaranteed jobs on board.

A *Dong Won 519* crewman called Supanto, who took the deal, was given US$9,800 and then had to pay $600 to the agent. He had claimed NZ$86,527 – US$74,500 – for the hours he had worked. He told people he had been ashamed to make the deal but was desperate and needed to feed his family. Another crewman, Komarudin, who had worked on *Dong Won 701*, received just US$4,744.35 despite being entitled to nearly ten times that amount.

Hasan told me the agents were preying on the men. 'They want the crew to be desperate and then they will sign the peace agreements.' The men had no wages and no jobs. They had believed, without really knowing why, that they would get justice in New Zealand. It was only as time had dragged on that they realised they would not. 'This is bad for the credibility of the New Zealand legal system,' Hasan said.

After I ran a story on this little operation, Sanford contacted me. It denied using manning agents and said it was up to the Employment Relations Authority, a government body that settles disputes, to determine the validity of the men's claims. Strictly speaking, Sanford was right: it did not use agents. Rather, it hired ships, and the ships' owners used agents.

This kind of smoke and mirrors was what had led the New Zealand government to announce it would end the business of foreign charter fishing boats in New Zealand waters. The bill requiring foreign vessels to be reflagged to New Zealand by 2016 had been tabled in parliament in February 2013. By early 2014 it was languishing well down the parliamentary order paper, unlikely to be passed before the general election in September.

However, publication across the country on April 8 of my interview with Hasan Nurhasan may have pricked some consciences: to my surprise on April 15 the bill was back in the House. The government had also decided to remove some of the exemptions, including the one under which Māori iwi were to have been given until 2020 to stop using slave boats.

Te Ururoa Flavell, the Māori Party's co-leader, asked in parliament whether iwi would be given compensation by the government for not being allowed to use foreign labour. 'Bearing in mind that the ability of Māori to move into the fisheries came out of an actual fisheries claim, can the minister tell the House how this policy does not breach, firstly, the Treaty of Waitangi, and, secondly, the Māori fisheries claim?'

Jo Goodhew, the associate primary industries minister, had strict instructions to brush off the complaints. The minister [for primary industries], she said, knew the law change would impact Māori 'but he is not prepared to allow behaviour that has been described as slavery at sea to continue in New Zealand waters. This bill is an essential step to stamp out this behaviour and to uphold New Zealand's international reputation.'

As the politicians debated the bill through its final stages, they would mostly claim credit for their own farsightedness and brush over the fact that such a monumental wrong had continued through successive governments.

I was not especially bothered by their failure to acknowledge the role the media had played in bringing about change. That is always the way. But what we and other like-minded men, women and organisations have achieved is a source of great pride. Today the appalling treatment of people, fish stocks and the world's oceans by many fishing companies is an increasingly urgent part of national and international conversations. It used to be thought that fishing had a kind purity about it – the harvesting of an excellent source of protein from pristine and ever-replenished waters. Nobody can say that any more. But nor can these companies and their practices any longer escape the notice of the world.

Acknowledgements

Now and again I was reminded of what was at stake in this book: a lot of money, none of it mine. Assorted insults and threats came my way. No one in the fishing industry is especially keen on transparency and open discussion about its dark side. I was occasionally unsure of whether phone calls were about fish or that I might join them. That said, I've decided to be cautious about to whom I offer public gratitude. Many wanted to help me understand a complex issue. They did not want a fight. They will recognise their contributions and, I hope, accept my thanks for their help and bravery.

Others I want to thank have outed themselves in the struggle. Paramount are Dr Glenn Simmons and Dr Christina Stringer of the University of Auckland Business School. Theirs was no ivory tower: they always knew the data they collected was about people and their families. In the struggle to end slave fishing, there were no others with such a clear vision of wrong and right.

One-time fisherman, master mariner and intense human rights advocate Daren Coulston never lost sight of what this story was about – and made darned sure I was not going to forget. Lawyers Peter Dawson in Nelson and Craig Tuck in Tauranga were always there on the phone, and put in far more non-billable hours for the exploited seamen than anybody else in their profession. They are a credit to the idea that we should help the oppressed and oppress the oppressors.

Peter Talley, chief executive of the Talley's Group seafood division, is tough enough to take my thanks on the chin. He was a crucial person in getting this story out. Alison Sykora of Sealord ably defended her company – and importantly let me have an insight into foreign fishing in New Zealand by allowing me to go to sea.

Thanks go to David Kemeys, editor of *The Sunday Star-Times* during the period my fishery stories appeared. David recognised an important story when he saw it. Fairfax Media tolerated my obsession. Journalist Ben Skinner of the Schuster Institute for Investigative Journalism in the United States was, in the beginning, competition on my beat, but throughout offered nothing but support and encouragement.

To Awa Press's Mary Varnham a special thanks; she believed even more than I thought I did. Her skill in persuading a weary journalist that a never-ending story might make a book should be retailed. It was powerful, persuasive and appreciated.

To my family and friends, who endured being around a difficult author, my thanks and apologies.

I hope this book makes a difference.

Abbreviations

ACC	Accident Compensation Corporation
AFL-CIO	American Federation of Labor and Congress of Industrial Organizations
AFP	Agence France-Presse
AIS	automatic identification system
CCAMLR	Commission for the Conservation of Antarctic Marine Living Resources
CITES	Convention on International Trade in Endangered Species of Wild Fauna and Flora
DSC	digital selective calling
EEZ	exclusive economic zone
EPIRB	electronic position-indicating radio beacon
FAD	fish aggregation device
FAO	Food and Agriculture Organization of the United Nations
FCV	foreign charter vessel
H&G	head and gut (of fish)
ICCAT	International Commission for the Conservation of Atlantic Tunas
ICIJ	International Consortium of Investigative Journalists
IMO	International Maritime Organization
IOM	International Organization for Migration
IUU	illegal, unregulated, unreported (fishing)
MARPOL	International Convention for the Prevention of Pollution from Ships 1973 and its protocol of 1978
MBIE	Ministry of Business, Innovation and Employment, New Zealand
MPI	Ministry for Primary Industries, New Zealand
MNZ	Maritime New Zealand
NATO	North Atlantic Treaty Organization
SOFIA	*State of World Fisheries and Aquaculture*
SPRFMO	South Pacific Regional Fisheries Management Organisation
STPP	sodium tripolyphosphate
UNESCO	United Nations Educational, Scientific and Cultural Organization
VHF	very high frequency (radio)
VMS	vessel monitoring system
WCPFC	Western and Central Pacific Fisheries Commission

Common and scientific names

abalone (*Haliotis asinina*)
Alaska pollock (*Theragra chalcogramma*)
albacore tuna (*Thunnus alalunga*)
anchoveta (*Engraulis ringens*)
Antarctic krill (*Euphausia superba*)
Antarctic toothfish (*Dissostichus mawsoni*)
Atlantic bluefin tuna (*Thunnus thynnus*)
Atlantic cod (*Gadus morhua*)
Atlantic herring (*Clupea harengus*)
Auckland Island arrow squid (*Nototodarus sloanii*)
barracouta (*Thyrsites atun*)
bigeye tuna (*Thunnus obesus*)
blue cod (*Parapercis colias*)
blue whiting (*Micromesistius poutassou*)
bluefin tuna (*Thunnus maccoyii*)
catfish (*Pangasius bocourti*)
Chinese bahaba (*Bahaba taipingensis*)
chub mackerel (*Scomber japonicus*)
cutthroat eel (*Synaphobranchoidei*)
common dolphin (*Delphinus delphis*)
electric ray (*Torpedo fairchildi*)
great white shark (*Carcharodon carcharias*)
hagfish (*Eptatretus cirrhatus*)
hake (*Merluccius bilinearis*)
halibut (*Reinhardtius hippoglossoides*)
hoki (*Macruronus novaezelandiae*)
humpback whale (*Megaptera novaeangliae*)
hāpuku (*Polyprion oxygeneios*)
jack mackerel (*Trachurus novaezelandiae*)
Japanese anchovy (*Engraulis japonicus*)
killer whale (*Orcinus orca*)
ling (*Genypterus blacodes*)
longnose velvet dogfish (*Centroscymnus crepidater*)
merluccid hake (*Merlucciidae*)

neon flying squid (*Ommastrephes bartramii*)
orange roughy (*Hoplostethus atlanticus*)
Pacific bluefin tuna (*Thunnus orientalis*)
Pacific salmon (*Oncorhynchus spp.*)
parrotfish (*Scarine labrids*)
Patagonian toothfish (*Dissostichus eleginoides*)
pāua (*Haliotis iris* and *Haliotis australis*)
pelagic kahawai (*Arripis trutta*)
rabbitfish (*Siganidae*)
red cod (*Pseudophycis bachus*)
rig (*Mustelus lenticulatus*)
Russian sockeye salmon (*Oncorhynchus nerka*)
sea lion (*Phocarctos hookeri*)
silverfish (*Pleuragramma antarcticum*)
skipjack tuna (*Katsuwonus pelamis*)
snake mackerel (*Gempylus serpens*)
snapper (*Chrysophrys auratus*)
southern bluefin tuna (*Thunnus maccoyii*)
southern blue whiting (*Micromesistius australis*)
southern right whale (*Eubalaena australis*)
squid (*Nototodarus gouldi* and *Nototodarus sloanii*)
striped marlin (*Kajikia audax*)
swordfish (*Xiphias gladius*)
totoaba (*Totoaba macdonaldi*)
trevally (*Pseudocaranx dentex*)
tuna and tuna-like species (*Scombroidei*)
white-capped albatross (*Thalassarche steadi*)
white-chinned petrel (*Procellaria aequinoctialis*)
wrasses (*Cheilinus undulatus*)
yellow-cyed penguin (*Megadyptes antipodes*)
yellowfin tuna (*Thunnus albacares*)

Endnotes

Author's note

1. Seafood New Zealand, 'Quota Management System (QMS)', http://www.seafoodnewzealand.org.nz/our-industry/sustainable-management/quota-management-system-qms/
2. New Zealand Ministry for Primary Industries, 'Quota Management System', December 21, 2010, http://fs.fish.govt.nz/Page.aspx?pk=81
3. New Zealand Legislation, Fisheries Act 1996, reprint as at 1 January 2014, http://www.legislation.govt.nz/act/public/1996/0088/latest/DLM394192.html
4. FishServe, 'Annual Catch Entitlement', https://www.fishserve.co.nz/information/annual-catch-entitlement, https://www.fishserve.co.nz/information/annual-catch-entitlement
5. New Zealand Legislation, Fisheries Act 1996, reprint as at 1 January 2014, http://www.legislation.govt.nz/act/public/1996/0088/latest/DLM394192.html

Preface

1. Information extrapolated from Fisheries and Aquaculture Department, Food and Agriculture Organization of the United Nations, *The State of World Fisheries and Aquaculture 2010*, http://www.fao.org/docrep/013/i1820e/i1820e.pdf, and *The State of World Fisheries and Aquaculture 2012*, *http://www.fao.org/docrep/016/i2727e/i2727e.pdf*
2. The best known example of species collapse is cod on the Grand Banks in Newfoundland in 1992: following the arrival of foreign factory trawlers, which began in the late 1950s, the fishery was wiped out and over 42,000 people lost their jobs. See Canada History, 'Cod collapse', http://www.canadahistory.com/sections/eras/pcsinpower/cod_collapse.htm and WWF Global, *Unsustainable Fishing, http://wwf.panda.org/about_our_earth/blue_planet/problems/problems_fishing/*. In the 1970s Peru's coastal anchovy fishery collapsed because of overfishing and El Nino weather. It has still not returned to levels once harvested. See US Library of Congress, *Fishing: Peru, http://countrystudies.us/peru/55.htm*. Sole fisheries in the Irish Sea and the west English Channel have also virtually collapsed. See Joint Nature Conservation Committee, 'The UK Biodiversity Action Plan 1992–2012', *http://jncc.defra.gov.uk/default.aspx?page=5155*. There are other examples such as bluefin tuna, now believed to be facing extinction.
3. Jane Nickerson, 'News of Food: Fish Sticks Soar in Public Favor With New Makers by Dozens in Field', *The New York Times*, May 20, 1954, 37, ProQuest

Historical Newspapers *The New York Times* (1851–2005) database document ID: 83334520; quoted by E. Robert Kinney in 'His Shtick was Fish Sticks', *AARP Blog*, http://blog.aarp.org/2013/05/17/e-robert-kinney-fish-sticks-inventor-gortons-seafood-co/

1 Dangerous seas

1. Information based on author's enquiries to seamen's missions, trade unions and diplomatic missions, while seeking the men's names and identities.
2. In May 2014, in the wake of extensive publicity in the United States about the failure of the sanctuary, Kiribati's president and cabinet finally voted to close all commercial fishing in the protected area by the end of that year. PIPSO: Pacific Islands Private Sector Organisation, 'Kiribati bans fishing in crucial marine sanctuary', May 2014, http://pipso.org/kiribati-bans-fishing-in-crucial-marine-sanctuary/

2 Frozen to death

1. Marine Casualty Investigation Team, Korean Maritime Safety Tribunal, *Investigation Report: The Sinking of the Fishing Vessel Insung 1*, October 2011, http://www.taic.org.nz/LinkClick.aspx?fileticket=9MTYH7KzmWo%3D&tabid=36&language=en-US
2. Ibid.
3. Insung Corporation, *Business E-Catalog*: http://www.insungnet.co.kr/eng/02business/01ecatal.php
4. *VietNamNet*, http://m.english.vietnamnet.vn/fms/society/2673/thousands-of-expatriate-sailors-face-risks-and-difficulties.html
5. Ibid.
6. Ibid.
7. Gareth Hughes MP, 'Ross Sea should be declared a Marine Protected Area', https://www.greens.org.nz/press-releases

3 Crime

1. Kimberly Warner, Walker Timme, Beth Lowell, Michael Hirschfield, 'Oceana Study Reveals Seafood Fraud Nationwide', February 2013, http://oceana.org/en/news-media/publications/reports/oceana-study-reveals-seafood-fraud-nationwide
   J.E. Foulke, 'Is something fishy going on? Intentional mislabelling of fish', *FDA Consumer*, September 1993.
2. United Nations Office on Drugs and Crime, 'Transnational Organized Crime in the Fishing Industry', April 13, 2011, https://www.unodc.org/documents/human-trafficking/Issue_Paper_-_TOC_in_the_Fishing_Industry.pdf
3. Environmental Justice Foundation, 'Bringing fishing vessels out of the shadows', http://ejfoundation.org/sites/default/files/public/EU_Global_Record_briefing_low-res-version_ok.pdf

4. United Nations Office on Drugs and Crime, Transnational Organized Crime in the Fishing Industry, April 13, 2011, https://www.unodc.org/documents/human-trafficking/Issue_Paper_-_TOC_in_the_Fishing_Industry.pdf

5. Martin Whitfield, *Out of Sight, Out of Mind: Seafarers, Fishers and Human Rights*, June 2006, International Transport Workers' Federation, http://www.itfseafarers.org/files/extranet/-1/2259/HumanRights.pdf

6. The case concerned Indonesian crew on board Korean boat *Sky 75*, see International Transport Workers' Federation, *Out of Sight, Out of Mind: Seafarers, Fishers and Human Rights*, June 2006, http://www.itfglobal.org/files/extranet/-1/2259/HumanRights.pdf.

7. United Nations Office for the Coordination of Humanitarian Affairs, 'Cambodia: Men being exploited, trafficked too', *Irin News*, http://www.irinnews.org/printreport.aspx?reportid=86155

8. 'United Nations Inter-Agency Project on Human Trafficking , 'Exploitation of Cambodian Men at Sea', http://www.ilo.org/wcmsp5/groups/public/---ed_norm/---declaration/documents/publication/wcms_143251.pdf

9. Brian O'Riordan, 'Growing Pains: Senegal: Child Labour', International Collective in Support of Fishworkers, http://www.icsf.net/en/samudra/detail/EN/2527.html

10. Quoted *in Ocean Revolution Moçambique*, http://www.state.gov/documents/organization/82902.pdf

11. Margot L. Stiles, Ariel Kagan, Emily Shaftel, Beth Lowell, *Stolen Seafood: The Impact of Pirate Fishing On Our Oceans*, Oceana, 2013, http://oceana.org/sites/default/files/reports/Oceana_StolenSeafood.pdf

12. 'Republic of Namibia, National Plan of Action to Prevent, Deter and Eliminate Illegal, Unreported and Unregulated Fishing', January 2007, http://209.88.21.36/opencms/export/sites/default/grnnet/MFMR/downloads/docs/Namibia_NPOA_IUU_Final.pdf

13. SADC Marine Fisheries Ministerial Conference to Stop Illegal Fishing: The Billion Dollar Treasure Hunt, media release, July 4, Namibia, http://www.stopillegalfishing.com/ministerial_conference.php

14. COLTO: Coalition of Legal Toothfish Operators, 'Rogues Gallery: The new face of IUU fishing for toothfish', October 2003, http://www.colto.org/wp-content/uploads/2013/01/Rogues-Gallery-Final.pdf

15. NSW Department of Primary Industries, 'Operation Fusion nets abalone trafficking ring', http://www.dpi.nsw.gov.au/archive/news-releases/fishing-and-aquaculture/2011/abalone-trafficking-ring

16. Press Association, *The Guardian*, 'Cockler gangmaster jailed for 14 years', http://www.theguardian.com/uk/2006/mar/28/ukcrime.world

17. Otto Bakano, 'A different kettle of fish', *iafrica.com*, http://news.iafrica.com/features/2207340.htm

18. Onislam and News Agencies, 'Somali Pirates Blessing to Kenya Fishermen', http://www.onislam.net/english/news/africa/438589.html

19. Havoscope: Global Black Market Information, http://www.havocscope.com/tag/illegal-fishing/
20. David Smith, 'Scourge of the Seas: Pirate Fishermen Plunder the World's Fish Supply', *Economy Watch*, May 9, 2013, http://www.economywatch.com/economy-business-and-finance-news/pirate-fishermen-plunder-the-ocean.09-05.html
21. Steve Trent, quoted in David Smith, 'Scourge of the Seas: Pirate Fishermen Plunder the World's Fish Supply', *Economy Watch*, May 9, 2013, http://www.economywatch.com/economy-business-and-finance-news/pirate-fishermen-plunder-the-ocean.09-05.html.

4 Oyang

1. Evidence reported by author at Coroners Court Wellington, April 16–20, 2012; full finding at http://www.justice.govt.nz/courts/coroners-court/media-centre/findings-of-public-interest/oyang-70-csu-2010-cch-000579
2. Ibid.

5 Southern blue whiting

1. Evidence reported by author at Coroners Court Wellington, April16–20, 2012; full finding at http://www.justice.govt.nz/courts/coroners-court/media-centre/findings-of-public-interest/oyang-70-csu-2010-cch-000579
2. See author's note on fisheries management, pages 1–3.
3. Evidence reported by author at Coroners Court Wellington, April 16–20, 2012; full finding at http://www.justice.govt.nz/courts/coroners-court/media-centre/findings-of-public-interest/oyang-70-csu-2010-cch-000579
4. Ibid.
5. Ibid.

6 Cover-up

1. Evidence reported by author at Coroners Court Wellington, April 16–20, 2012; full finding at http://www.justice.govt.nz/courts/coroners-court/media-centre/findings-of-public-interest/oyang-70-csu-2010-cch-000579
2. Ibid.
3. Interview with author, name withheld.
4. Evidence reported by author at Coroners Court Wellington, April 16–20, 2012; full finding at http://www.justice.govt.nz/courts/coroners-court/media-centre/findings-of-public-interest/oyang-70-csu-2010-cch-000579
5. Ibid.

7 Slavery

1. Author, 'Slavery at sea', *The Sunday Star-Times*, April 3, 2011.
2. Ibid.
3. Leaked email between Sajo Oyang Corporation and Southern Storm Fishing Ltd.

4. Email to author from Tony Sung, legal/trade adviser, Embassy of the Republic of Korea to New Zealand, Wellington, March 9, 2012.
5. Christina Stringer, Glenn Simmons and Daren Coulston, *Not in New Zealand Waters, Surely? Labour and human rights abuses aboard foreign fishing vessels*, New Zealand Asia Institute Working Paper Series, September 15, 2011, http://docs.business.auckland.ac.nz/Doc/11-01-Not-in-New-Zealand-waters-surely-NZAI-Working-Paper-Sept-2011.pdf
6. John Albert Situmeang, email to author, December 6, 2010.
7. Christina Stringer, interview with author.

8 Verdict

1. Findings of Coroner R.G. McElrea, Inquiry into the death of Yuniarto Heru, Samsuri, Taefur, Wellington, March 6, 2013, http://www.justice.govt.nz/courts/coroners-court/media-centre/findings-of-public-interest/oyang-70-csu-2010-cch-000579
2. Ibid.
3. Ibid.
4. Ibid.
5. Ibid.

9 Tangaroa's bounty

1. Captain James Cook, log entry, October 15, 1769, quoted *in Report of the Waitangi Tribunal on the Muriwhenua Fishing Claim*, http://archive.is/XndL#selection-265.0-265.63
2. Joseph Banks, quoted in *Report of the Waitangi Tribunal on the Muriwhenua Fishing Claim*, http://archive.is/XndL#selection-265.0-265.63
3. William Colenso, quoted in *Report of the Waitangi Tribunal on the Muriwhenua Fishing Claim*, http://archive.is/XndL#selection-265.0-265.63
4. *New Zealand Journal*, London, 1843, quoted in *Report of the Waitangi Tribunal on the Muriwhenua Fishing Claim*, http://archive.is/XndL#selection-265.0-265.63.
5. Āpihai Te Kawau. Ngāti Whātua, quoted in *Report of the Waitangi Tribunal on the Muriwhenua Fishing Claim*, http://archive.is/XndL#selection-265.0-265.63.
6. Waitangi Tribunal, Te Rōpū Whakamana i Te Tiriti o Waitangi, *Report of the Waitangi Tribunal on the Muriwhenua Fishing Claim*, http://archive.is/XndL#selection-265.0-265.63.
7. Waitangi Tribunal Te Rōpū Whakamana i Te Tiriti o Waitangi, *The Ngāi Tahu Sea Fisheries Report 1992*, https://forms.justice.govt.nz/search/Documents/WT/wt_DOC_68472628/NT%20Sea%20Fisheries%20W.pdf
8. Aramanu Ropiha, email to author.
9. Pita Sharples, Māori Affairs minister, reported by author in 'Māori warned over ageing foreign fishing boats, *The Sunday Star-Times*, May 25, 2011, http://www.stuff.co.nz/national/5051905/Maori-warned-over-ageing-foreign-fishing-boats

10. New Zealand Government media release, 'Ngati Tama Signs Heads of Agreement', September 24, 2009, at http://www.reocities.com/ken_sims_98/ nzffa/tama.htm

11. Quoted in 'Time to lance the boil in fishing industry shame', Maritime Union of New Zealand, July 14, 2011, http://www.munz.org.nz/2011/07/14/time-to-lance-the-boil-in-fishing-industry-shame/

12. *Professional Skipper Magazine*, VIP Publications, Auckland, July/August 2011.

13. Author interview, name withheld.

14. Under New Zealand admiralty law, a party can issue proceedings directly against a vessel if it has a claim relating to the vessel, and the vessel can be arrested and held until the claim is settled. If it is not settled, the party can direct the court to sell the vessel, with the proceeds being used to settle the claim, see http://www. maritimelaw.org.nz/adm73.html

15. Yvonne Tahana, 'Taranaki iwi loses almost $20m', *The New Zealand Herald*, April 18, 2012, http://www.nzherald.co.nz/nz/news/article.cfm?c_ id=1&objectid=10799666

16. Juliet Smith, 'One man's journey to justice; History of Ngāti Tama claim', *Taranaki Daily News*, February 23, 2002, http://www.ngatitama.net/upload/ One%20Man%27s%20Journey%20to%20Justice%20-%20History%20of%20 the%20Ngati%20Tama%20Claim.pdf

17. Glenn Simmons and Christina Stringer, 'New Zealand's Fisheries Management System: Forced Labour an Ignored or Overlooked Dimension?', New Zealand Asia Institute, *Marine Policy*, 2014 (forthcoming).

18. Ministry for Primary Industries, correspondence obtained by author under Official Information Act.

10 Pirate ship

1. Captain Paul Watson, 'A Kiwi Corporate Whore in the Land of the Rising Sun', Sea Shepherd Conservation Society, February 9, 2009, http://www.seashepherd. fr/news-and-media/editorial-090209-1.html

2. Glenn Inwood, quoted by author in 'Families of fishing crew face backlash', Fairfax Media, August 10, 2011, http://www.stuff.co.nz/national/5261854/ Families-of-fishing-crew-face-backlash

3. Glenn Inwood, quoted by author in '$10,500 fine for fishing boat's secret dumping', Fairfax Media, February 22, 2013, http://www.stuff.co.nz/ business/8340596/10-500-fine-for-fishing-boats-secret-dumping

4. *Abstracts Business, International*, '$1.8m fine on Spanish owner – while another mounts challenge', http://www.readabstracts.com/Business-international/18m-fine-on-Spanish-owner-while-another-mounts-court-challenge-.html

5. Leaked email between Sajo Oyang Corporation and Southern Storm Fishing Ltd.

6. David Clarkson, 'Oyang 75 dumped up to $1.4m of fish – Fisheries', *The Press*, July 19, 2012, http://www.stuff.co.nz/the-press/news/7129679/Oyang-75-dumped-up-to-1-4m-of-fish-Fisheries

7. Judge D. J. L. Saunders, judgement, Ministry of Fisheries v Chong Pil Yun and others, District Court Christchurch, June 22, 2012, CRI-2001-009-011297.
8. National Human Rights Commission of Korea, statement to Fairfax Media, May 2012.
9. PricewaterhouseCooper, report for New Zealand Department of Labour, obtained by author under Official Information Act.
10. Yoo Kyungpil, Busan regional public prosecutor, Republic of Korea, report, information obtained by author from sources close to investigation.
11. From leaked email between Sajo Oyang Corporation and Southern Storm Fishing Ltd.
12. Author, 'Key to negotiate South Korean free trade deal', *Fairfax NZ News*, March 23, 2012, http://www.stuff.co.nz/national/politics/6625606/Key-to-negotiate-South-Korean-free-trade-deal

11 The politics of fishing
1. Woods Hole Oceanographic Institution, 'What are Seamounts?', http://www.whoi.edu/main/topic/seamounts
2. Whaimutu Dewes, quoted in 'Argentinian snag for Sealord', Fairfax Media, July 7, 2013, http://www.stuff.co.nz/business/8884273/Argentinian-snag-for-Sealord
3. Tracy Watkins, 'Cosgrove "upfront" over mate's donation', *The Dominion Post*, July 9, 2012, http://www.stuff.co.nz/dominion-post/news/politics/7242978/Cosgrove-upfront-over-mates-donation
4. Clayton Cosgrove, author interview.
5. Shane Jones, author interview.

12 Aboard a Soviet trawler
1. Rebecca Surtees, *Trafficked at sea. The exploitation of Ukrainian seafarers and fishers 2012*, International Organization for Migration, Geneva, and NEXUS Institute, Washington D.C., 2013, http://www.nexusinstitute.net/publications/pdfs/Trafficked%20at%20sea%20web.pdf

13 Inquiry
1. Soon Nam Oh, Southern Storm Fishing Ltd, submission to Ministerial Inquiry into the Use and Operation of Foreign Charter Vessels, Ministry for Primary Industries, February 2012, http://www.fish.govt.nz/NR/rdonlyres/A15D995C-A00D-4DA6-A3D9-04B5849E734D/0/Vol5.pdf
2. Dawson & Associates, Submission to Ministerial Inquiry into the Use and Operation of Foreign Charter Vessels, October 6, 2011, http://www.fish.govt.nz/NR/rdonlyres/A603543E-606B-4BF0-B7F8-64F2242CFECA/0/Vol2.pdf
3. Talley's Group Ltd, Submission to Ministerial Inquiry into the Use and Operation of Foreign Charter Vessels, September 2011, http://www.fish.govt.nz/NR/rdonlyres/90CFB8CC-C98C-42CB-A7C5-30A96AEE09E2/0/Talley_FCV_Submission_vol15a.pdf

4. Office of the Minister for Primary Industries and Office of the Minister of Labour, 'Government Response to the Ministerial Inquiry on Foreign Charter Vessels' (redacted), undated, http://www.fish.govt. nz/NR/rdonlyres/7EB13F69-B081-4B44-A298-2BAE1C214B26/0/ govtresponsetotheministerialinquiryonforeigncharvervessels.pdf
5. Author, '$300 million at stake if FCVs forced out', *Fairfax NZ News*, July 31, 2013, http://www.stuff.co.nz/business/industries/8986505/300-million-at-stake-if-FCVs-forced-out

14 Melilla

1. E. Benjamin Skinner, author interview.
2. E. Benjamin Skinner, 'The Fishing Industry's Cruelest Catch', *Bloomberg Businessweek*, February 23, 2012, http://www.businessweek.com/ articles/2012-02-23/the-fishing-industrys-cruelest-catch
3. Ibid.
4. Ibid.

15 Sanford

1. Quoted by Alliance for American Manufacturing in 'Sorry Charlie? American-made tuna claim questioned', March 10, 2013, http://americanmanufacturing. org/blog/sorry-charlie-american-made-tuna-claim-questioned
2. International Maritime Organization, International Convention for the Prevention of Pollution from Ships (MARPOL), entry into force October 2, 1883, http://www.imo.org/About/Conventions/ListOfConventions/Pages/ International-Convention-for-the-Prevention-of-Pollution-from-Ships-(MARPOL).aspx
3. From documents filed in US District Court District of Columbia, United States of America v Sanford Ltd, Roland Ong Vano and James Pogue, Criminal No. 1:11-cr-00352-BAH, May 4, 2012.
4. Sanford Ltd, 'Sanford Relies on Clean Ocean and Rejects US Government Allegations of Pollution and Obstruction', press release, December 7, 2011, http://www.sanford.co.nz/shadomx/apps/fms/fmsdownload.cfm?file_ uuid=0E1A28B4-1FB6-4050-90D9-7DF775686355&siteName–sanfordfisheries
5. From documents filed in US District Court District of Columbia, United States of America v Sanford Ltd, Roland Ong Vano and James Pogue, Criminal No. 1:11-cr-00352-BAH, May 4, 2012.
6. Fili Sagapolutele, *The Samoa News*, '$2.4 Million Fine for American Samoa Oil Dumping', *Pacific Islands Report*, January 12, 2013, http://pidp.org/ archive/2013/January/01-14-04.htm
7. Sanford Ltd, *Annual Report 2003*, http://www.sanford.co.nz/sanfordfisheries/ fms/documents/Annual_Report_2003.pdf
8. Information leaked to author from sources close to investigation.

16 The Southern Ocean

1. AAP, 'Viarsa captain praises justice system', November 7, 2005, http://archive.is/Wnkbu#selection-977.0-1450.0
2. The Center for Public Integrity and The International Consortium of Investigative Journalists, *Looting the Seas II,* http://ftp.trojkan.se/temp/Reportage/Fredrik%20Laurin/looting%20of%20the%20seas_2.pdf
3. Andrew Darby, 'Pirate boats slip legal nets to fish with impunity', *Sydney Morning Herald*, March 5, 2005, http://www.smh.com.au/news/National/Pirate-boats-slip-legal-nets-to-fish-with-impunity/2005/03/04/1109700681878.html
4. Felicity Wong, quoted in 'Flying our flag to protect our patch', *The Evening Post*, April 17, 2000.
5. Ibid.
6. Colin James, 'Fishing pirates slice through seas of toothless treaties', *The New Zealand Herald*, May 4, 2004, http://www.nzherald.co.nz/business/news/article.cfm?c_id=3&objectid=3552729
7. Omunkete Fishing (Pty) Ltd v Minister of Fisheries, High Court, Wellington, CIV 2008-485-1310, http://www.maritimelaw.org.nz/0508.html
8. Author, 'Fishing firms lied about catches', *Fairfax NZ News*, February 10, 2014, http://m.stuff.co.nz/national/9705332/Fishing-firms-lied-about-catches. Documentation at http://www.ccamlr.org
9. Ko Na-mu, 'Korean fishing vessels accused of sex crimes in international waters', *The Hankyoreh*, May 12, 2012, http://english.hani.co.kr/arti/english_edition/e_international/532524.html%20
10. Margot L. Stiles, Ariel Kagan, Emily Shaftel, Beth Lowell, *Stolen Seafood: The Impact of Pirate Fishing on Our Oceans*, Oceana, http://oceana.org/sites/default/files/reports/Oceana_StolenSeafood.pdf
11. Clive Evans, author interview.
12. David Ainley, author interview.
13. Ibid.
14. Author, 'NZ fishing cuts whale numbers in Antarctica', March 19, 2011, *Fairfax NZ News*, http://www.stuff.co.nz/national/4788118/NZ-fishing-cuts-whale-numbers-in-Antarctica
15. Interpol distributes purple notices 'to provide information on modus operandi, procedures, objects, devices and concealment methods used by criminals'. http://www.interpol.int/INTERPOL-expertise/Notices/Purple-notices-%E2%80%93-public-versions

17 The Pacific prize

1. Eugene Bingham, 'Swallowed by the Sea', August 16, 2003, *The New Zealand Herald* (print edition only).
2. South Pacific Forum, Convention for the Prohibition of Fishing with Long Driftnets in the South Pacific (1989), http://eelink.net/~asilwildlife/southpacific.html

3. Sea Shepherd, 'Death trawler on its way Down Under', August 7, 2012, http://www.seashepherd.org/news-and-media/2012/08/07/death-trawler-on-its-way-down-under-1418

4. François Doumenge, *The social and economic effects of tuna fishing in the South Pacific*, South Pacific Commission, New Caledonia, 1966.

5. Craig Van Note, 'Dolphins still die for druglords', International Marine Mammal Project, June 25, 2009, http://earthisland.org/marinemammal/index.php/eco2009/eco2009madeira/dolphins-still-die-for-druglords/

6. WWF, 'Southwest Pacific Longline Caught Albacore: Going, Going, Gone?, Prepared for the Western and Central Pacific Fisheries Commission Meeting, Guam, March 25–29, 2012, https://www.wwf.or.jp/activities/upfiles/WWF20120325SPA-PolicyBrief.pdf

7. Mark Price, 'Fishy business', *Otago Daily Times*, June 23, 2012, http://www.odt.co.nz/lifestyle/magazine/214296/fishy-business

8. Ibid.

18 The China syndrome

1. Gidget Fuentes, 'New carrier role in Pacific: fight illegal fishing', *Navy Times*, June 21, 2012, http://www.navytimes.com/article/20120621/NEWS/206210323/New-carrier-role-Pacific-fight-illegal-fishing

2. Ibid.

3. Michael Lodge, author interview.

4. Papers for the Preparatory Conference for the Commission for the Conservation and Management of Highly Migratory Fish Stocks in the Western and Central Pacific, December 6–7, 2005, Pohnpei, Micronesia, WCPFC/PrepCon/DP.36.

5. Roland Blomeyer, Ian Goulding, Daniel Pauly, Antonio San and Kim Stobberup, 'The Role of China in World Fisheries', Directorate General for Internal Policies, Policy Department B: Structural and Cohesion Policies, Fisheries, European Parliament, http://www.europarl.europa.eu/meetdocs/2009_2014/documents/pech/dv/chi/china.pdf

6. United States Congress, 'China's global quest for resources and implications for the United States', Hearing before the US-China Economic and Security Review Commission, January 26, 2012, http://origin.www.uscc.gov/sites/default/files/transcripts/1.26.12HearingTranscript.pdf

7. Russell Dunham, author interview.

8. Details of Chinese fishing subsidies in 'Fisheries Subsidies in China', The Pew Charitable Trusts Sea Around Us Project, Fisheries, Ecosystems and Biodiversity, http://www.seaaroundus.org/Subsidy/default.aspx?GeoEntityID=37

9. Christina Stringer, Glenn Simmons, Eugene Rees, 'Shifting post production patterns: exploring changes in New Zealand's offshore processing', *New Zealand Geographer*, 67, 2011.

ENDNOTES   227

10. Oceana, *IUU Fishing: Overview*, http://oceana.org/en/eu/our-work/responsible-fishing/dirty-fishing/iuu-fishing/overview
11. Molly McGrath, Molly, 'The True Cost of Shrimp', Solidarity Center, Washington, D.C., http://www.academia.edu/1580285/The_True_Cost_of_Shrimp_2008_Solidarity_Center_
12. David Smith, 'Chinese appetite for shark fin soup devastating Mozambique coastline', *The Guardian*, February 14, 2013, http://www.theguardian.com/world/2013/feb/14/chinese-shark-fin-soup-mozambique
13. Simon Parry, 'Air Pacific accused of hypocrisy over shark fins cargo', *South China Morning Post*, May 15, 2013, http://www.scmp.com/news/hong-kong/article/1237843/air-pacific-accused-hypocrisy-over-shark-fins-cargo
14. Ibid.
15. Alex Hofford, author interview.
16. Census of Marine Life, 'Effects of shark decline', 2009, http://www.coml.org/discoveries/trends/shark_decline_effects

19 The Russians are coming

1. John Amos, author interview.
2. *Carmen Tere* crew member, author interview.
3. Ibid.
4. Spokesperson for Angelsonia Pesca, author interview.
5. South Pacific Regional Fisheries Management Organisation, documents at https://www.southpacificrfmo.org
6. Ibid.
7. In proceedings conducted by The Review Panel established under Article 17 and Annex II of the Convention on the Conservation and Management of High Seas Fishery Resources in the South Pacific Ocean with regard to The Objection by the Russian Federation to a Decision of the Commission of the South Pacific Regional Fisheries Management Organisation, Findings and Recommendations of the Review Panel, July 5, 2013, The Hague, The Netherlands, http://www.google.co.nz/url?sa=t&rct=j&q=&esrc=s&source=web&cd=2&ved=0CC8QFjAB&url=http%3A%2F%2Fwww.pca-cpa.org%2Fshowfile.asp%3Ffil_id%3D2289&ei=BdRmU4-qDIXskAXhnoGgCA&usg=AFQjCNHWjZNIdTQZyDPUgjX3byyCCtB6_g&bvm=bv.65788261,d.dGI

20 Wars

1. Food and Agriculture Organization of the United Nations, Fisheries and Aquaculture Department, *The State of the World Fisheries and Aquaculture (SOFIA)*, http://www.fao.org/fishery/sofia/en
2. CITES, 'Governments not ready for trade ban on bluefin tuna', March 18, 2010, http://www.cites.org/eng/news/pr/2010/20100318_tuna.shtml
3. Adam Gabbatt and agencies, 'EU backing for bluefin tuna trade ban sparks Japan protests', *The Guardian*, March 11, 2010, http://www.theguardian.com/

world/2010/mar/11/bluefish-tuna-ban-japan-protests

4. Ibid.

5. Sebastián Losada, *Pirate Booty: How ICCAT is failing to curb IUU fishing*, Greenpeace Spain, September 2007, http://www.imcsnet.org/imcs/docs/how_ iccat_failing_curb_iuu.pdf

6. Asser Institute report on court findings, T-367/10, Bloufin Touna Ellas Naftiki Etaireia a.o. v Commission, February 27, 2013, http://www.asser.nl/default. aspx?site_id=7&level1=12218&level2=12247&level3=12487

7. Dan Charles, 'Making Food From Flies (It's Not That Icky)', National Public Radio, September 19, 2013, http://www.npr.org/blogs/ thesalt/2013/09/19/223728061/making-food-from-flies-its-not-that-icky

21 Hope

1. William J. Broad, 'From Deep Pacific, Ugly and Tasty, With a Catch', *The New York Times*, September 9, 2009, http://www.nytimes.com/2009/09/10/ science/10fish.html

2. Sarah Crystell, author interview.

3. Ibid.

4. Jim McCarthy, author interview.

5. Ibid.

6. Author, 'Tuna Tactics: The buccaneering world of high seas fishing', http:// forums.allcoast.com/89-long-range-fishing-reports-discussion/14996-yft- lifecycle-question-2.html

7. Sonny Tau, author interview.

8. The Pew Charitable Trusts, Environmental Initiatives, 'Fish Aggregating Devices (FADS) Position Paper', June 28, 2011, 'http://www.pewenvironment. org/news-room/other-resources/fish-aggregating-devices-fads-position- paper-85899361234/en-EN

9. Quoted in every edition of *IHS Fairplay*, http://www.fairplay.co.uk

Index

living conditions *see* working
conditions
Lloyd's Register 57, 68, 205
Lodge, Michael 174, 201–202
longlining 9–10, 131, 169, 175, 203
Loung Dar 161, 162
Louwrens, Bheema 112, 114–116
Lung Yuin 174
Lyall, Greg 50, 53–54, 69

Macdonald, Ian 151
Macfarlane, Alastair 124
mackerel 79, 87, 112, 113–114,
165–167, 189, 190, 193, 195, 197
MacLaren, Grahame 83
Malaysia 19, 25, 26
Mallory, Tabitha Grace 176, 177
Mamaloni, Solomon 23
Mansfield, Bill 166
Māori 2, 45, 75–83, 84–85, 86–87, 89,
104–105, 125–126, 202, 211
See also individual iwi; quotas,
trading; Treaty of Waitangi
Māori Fisheries Trust, The *see* Te Ohu
Kaimoana
Māori Party 81–82, 211
Marcia 707 27
Margiris 166–167
marine protected areas 10 11, 51, 160
Marine Stewardship Council 174, 201,
204
Maritime New Zealand 100, 144, 162
inspection of vessels 37, 68, 71–72,
99–100
Rescue Coordination Centre 53
Maritime Union of New Zealand 83
MARPOL 137, 138
Matiu, McCully 78
Matuku, Wiremu 84–85

Mazzetta Company 142, 143
Mazzetta, Tom 143
McCarthy, Jim 201
McDonald's 200, 201
Filet-O-Fish 5–6, 179, 199
McElrea, Richard 56, 57, 67–68, 69,
70–71, 72–73, 74
McGrath, Sarah 121
Melilla 201 127–128, 129–132
Melilla 203 127–128, 129–132
Ministry for Primary Industries 38,
86–87, 100, 112, 116, 129–130, 187
Ministry of Fisheries 38, 65, 91, 92,
154–155
observers 38, 39, 40–41, 68, 72–73,
91, 93–94, 112, 169–170
See also Ministry for Primary
Industries
Moon, Kyung-so 62–63
Morecambe Bay 30–31
Morin, Mark 173
Moriori 82
Mozambique 29, 181, 197
Muriwhenua 77–78
mussels 183, 105

Namibia 29, 151, 154
Navy Times 173
New York Times, The 8, 199, 200–201
New Zealand
Accident Compensation Corporation
57, 58, 64
Air Force (RNZAF) 11, 12, 20, 50,
133, 152, 154, 155, 157, 162
Companies Office 62
Department of Labour 39, 92, 96,
129, 130
early history 75–77, 78, 79, 82,
140–141

The most important book about the environment in years
— Barbara Demick

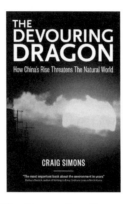

The Devouring Dragon
Craig Simons

The horrifying pollution within China is well known. Not so well known is the environmental impact of the country's rapid economic growth on the rest of the planet. In a few short years China has become the major market for endangered wildlife, the leading importer of tropical trees, and the biggest emitter of greenhouse gases. In this brilliant book, Craig Simons visits many of the places and people around the world affected by China's insatiable demand for resources and produces an unforgettable report.

'Craig Simons goes looking for the collateral damage of China's economic miracle and finds it in some surprising places, skillfully marshalling the small vignettes that tell the big story'
The New Zealand Herald

'A gripping new assessment of China's rise'
South China Morning Post

'Adroitly blending science with personal stories and on-the-road reportage, this book vividly describes the risks posed by climate change and biodiversity loss as Chinese consumers follow the unsustainable path set by their counterparts in the West'
Jonathan Watts, author of *When a Billion Chinese Jump*

'China's rise, as Simons shows, has raised hundreds of millions of its people out of poverty; the cost, environmentally and ecologically, has been calamitous'
Edward A. Gargan, author of *The River's Tale: A Year on the Mekong*

Available from all good bookstores and online at
awapress.com

The dramatic story of a disaster that shook a nation — and the world

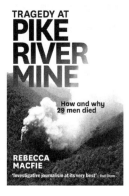

Tragedy at Pike River Mine
Rebecca Macfie

On a sunny afternoon in November 2010, a massive explosion rocked an underground coal mine deep in a New Zealand mountain range. A hundred and one minutes later two ashen men stumbled from the mine's entrance. Twenty-nine men remained trapped inside. For five agonising days their families and friends waited and prayed, until finally all hope was extinguished.

Pike River Mine had been touted as a showcase of modern mining, and shares in the company had been rapidly taken up by investors. Beneath the hype, though, lay mismanagement, mistakes and wilful blindness that would cost men their lives. Based on extensive research and over a hundred interviews, this powerful book provides chilling insights into the causes of the tragedy, and puts a human face on the people who suffered, and suffer still.

'Rebecca Macfie's spare, forceful narrative will leave readers far wiser and deeply outraged' Rod Oram, business journalist

'A devastating account of a needless tragedy – a must-read for all those who find it incomprehensible that such an event can occur in the 21st century in a supposedly developed, modern nation'
Victor Billot, *Otago Daily Times*

'Riveting and extraordinarily authoritative'
Cate Brett, former editor, *The Sunday Star-Times*

'Brilliant, hard to put down'
Duncan Garner, RadioLive